TESTING DIGITAL CIRCUITS
an introduction

ASPECTS OF
INFORMATION TECHNOLOGY

This series is aimed primarily at final year undergraduate and postgraduate students of Electronics and Computer Science, and provides an introduction to research topics in Information Technology which are currently being translated into teaching course material. The series aims to build bridges between foundation material covered in the first two years of undergraduate courses and the major research topics now attracting interest within the field of IT.

The format of the series is deliberately different from that of typical research reference works within the fields of interest. Depth of coverage is restricted in favour of providing a readable and comprehensible introduction to each topic, and to keep costs within the requirements for a course textbook. Nevertheless, each book provides a comprehensive overview and introduction to its subject, aimed at conveying the key elements in an attractive and clear fashion.

Each chapter is terminated by a summary itemizing the key points contained within, and problems and exercises are provided where appropriate to enable the student to test his knowledge. For the serious student, each book contains a comprehensive further reading list of texts and key reference papers covering the field, as well as giving an indication of those journals which publish within the area.

Series editors: **A C Downton** *University of Essex*
R D Dowsing *University of East Anglia*

TESTING DIGITAL CIRCUITS

an introduction

B.R. Wilkins
Senior Lecturer in Electronics
Southampton University

 Van Nostrand Reinhold (UK) Co. Ltd

First published in 1986 by
Van Nostrand Reinhold (UK) Co. Ltd
Molly Millars Lane, Wokingham,
Berkshire, England

Typeset in 10/12 pt Times by
Colset Private Ltd, Singapore

Printed in Great Britain by
The Thetford Press Ltd,
Thetford, Norfolk

British Library Cataloguing in Publication Data

Wilkins, B.R.
    Testing digital circuits: an
    introduction.
    1. Digital electronics—Testing
    I. Title
    621.3815′3   TK7868.D5

    ISBN 0–442–31748–4

CONTENTS

PREFACE

There can be few undergraduate courses in electrical or electronic engineering that do not include as a major component a study of the analysis and design of digital circuits. With the dramatic improvements in the cost and capabilities of integrated circuits that have taken place in the last couple of decades this subject is rightly regarded as being of central importance. It is not, however, enough just to design a circuit; after it is built it has to be tested.

This book is intended partly to fulfil a missionary function: there may still be engineers or managers in industry who have yet to be convinced of the significance of testing problems, and who, as a result, display some reluctance to make concessions to testability if they involve circuit overheads apparently resulting in higher costs of production. One important message of this book is that the economic arguments are much more complicated than this, and that judicious expenditure in the design stages will result in overall reductions in cost.

This economic consideration leads to a second, and perhaps more important, missionary function of the book: if good practical circuit design needs to take account of testability, it follows that testing ought to appear as an integral part of every undergraduate course in digital system design. The importance of the subject is gradually being recognized, but there are still far too many courses that pay it little or no attention. Part of the reason for this neglect might be the lack of appropriate text-books: I certainly felt this lack when preparing courses at Southampton University, and this provided one motivation for writing this book. Another part of the reason could be that the subject lacks academic respectability: there is little in the way of tidy mathematical procedures yielding unique solutions to problems. Again I am painfully aware of this feature, especially when attempting to formulate examination questions! However, I am convinced that, in engineering at least, any subject must earn its place in the academic curriculum by virtue of its practical importance, irrespective of whether it possesses theoretical elegance; on this basis, testing should come high on the list.

This book is intended to provide material for an introductory course in the subject. To follow this material the student requires very little in the way of previous knowledge: an understanding of basic Boolean algebra (including an appreciation of Karnaugh maps and state transition diagrams) and some acquaintance with standard digital components (such as the 7400 series of TTL SSI chips) will provide all the background that is necessary.

Chapter 1 attempts to set the scene first by discussing where testing fits into the manufacturing process, and then by indicating some of the economic considerations that come into play. If students can once be convinced that the problem exists, a major barrier to progress will have been removed.

Chapters 2 and 3 deal with the development of structurally-based test sets for combinational circuits using sensitive path techniques. These methods, or variations of them, provide the basis for all existing automatic test pattern generation systems: methods based on theoretical constructs (such as Boolean difference) are computationally impractical for circuits of realistic size.

Chapter 4, in discussing the problems of fault diagnosis, introduces some of the underlying theory of signature analysers based on LFSRs. There is no doubt that data compression techniques are essential not only in diagnostic probing applications, but also in self-testing systems. The feedback shift register has been widely adopted for these purposes: the linear theory presented here provides an introduction to a more comprehensive study which would include consideration of registers with non-linear feedback.

The main aim of Chapter 5, in introducing some of the major problems involved in testing sequential circuits, is to demonstrate the way in which quite simple circuits can be difficult or impossible to test. This point does need to be demonstrated: it is part of the missionary function mentioned earlier. Before designers can be expected to accept the constraints of 'design for testability' they must first be convinced that unconstrained design is unacceptable.

After a brief consideration (in Chapter 6) of some of the special problems associated with particular LSI devices, attention is then turned in Chapters 7 and 8 to perhaps the most important theme of the book: the various ways in which design needs to be influenced if testing problems are to be kept within bounds. The use of some or all of these techniques is incorporated into the in-house design rules imposed by individual companies, so that every newly-emergent graduate should at least be aware of the principles and the reasons for adopting them. It is equally important to appreciate the possibility of self-testing, which is likely to become of increasing importance in the next few years: this subject is introduced in Chapter 9.

The final chapter provides an introduction to the literature of the subject, which will have to be explored by any serious student. Although it is certainly useful to be provided with a selection of key references, I personally feel that references within the body of a tutorial text are somewhat distracting, and I suspect that most students generally ignore them. By gathering them together in a separate chapter, I hope that the usefulness will be preserved and the distraction avoided.

A few sample questions have been provided at the end of each chapter by way of exercises, and at the end of the book some notes have been included to suggest ways in which these questions might be answered. In a largely non-analytical subject it is impossible to cover the material adequately just using numerical questions with unique solutions. Among the exercises, therefore, there are some questions requiring descriptive answers and others for which many alternative answers are possible.

Finally, I must acknowedge my indebtedness to many people who have helped me to bring this material together. It is doubtful whether one person alone could ever expect to write a technical book unaided. This is particularly so when the person concerned is an academic, and the book deals with an essentially practical subject. To try and separate the important problems from the unimportant, and to establish the distinctions between theoretical solutions and practical procedures, I have been heavily dependent on consultations with many individuals in industry. I would like to record my thanks in particular to Ben Bennetts and his colleagues at Cirrus Computers, who first introduced me to the subject and allowed me to practise with their equipment; Alex Elliott of British Telecom, David Balston of Plessey Electronic Systems Research, and Ian Pearson of Inmos, all of whom have provided research support; Don Murray and his colleagues at ICL, Alan Boyce of Marconi Research, Peter Tullett of Plessey Marine, and John Ibbotson of IBM, who have provided invaluable help and advice; and to many undergraduate and postgraduate students who have by their questions forced me to think about the details of the subject. Similarly, a number of people, including several of those listed above, and, especially, Andy Downton and Bob Damper of Southampton University, have looked at parts of the manuscript and have provided helpful comments and suggestions and enabled me to remove some errors and ambiguities. For the errors that remain (and it would be foolish to suppose that there are none) I must accept responsibility: comments from readers would be most gratefully received.

1

TESTING IN CONTEXT

1.1 THE NEED FOR TESTING

In one way or another, the need for testing has always been with us. Ever since manufactured items were first produced, it has been generally accepted without question that, before the item is released to the customer, the manufacturer should establish that it has been assembled correctly and is working as it was intended to do. For a long time, and for very many products, the implementation of this concept has not been seen to present any particular problem; test and inspection procedures could be informal, based on an intuitive understanding of the product by the quality control department which applied a final inspection at the end of the production line. Such an approach is entirely appropriate to simple devices; any reader could easily compile a check-list for a wheelbarrow (e.g. Are the axle-nuts secure? Does the wheel turn freely? Are the legs splayed sufficiently to make the barrow stable? Are the hand-grips securely fitted? and so on), and the likelihood is that an inspection based on such a list would be sufficient to guard against an unsatisfactory item leaving the factory. Furthermore, even faults that were not envisaged by the compiler of the check list would, in most cases, be detected simply on the basis of a visual inspection and the common sense of the inspector.

Electronic systems present rather more difficulties, if only because visual inspection clearly cannot cover the most important facets of the system's performance. Even so, for a simple system such as a radio, additional performance criteria can be suggested on an intuitive basis; at its crudest, perhaps just a subjective assessment of sound quality obtained from one station on each waveband. This could be augmented with objective measurements such as frequency responses, but even inspection based on the simplest criteria is likely to provide a reasonable assurance that the unit is free from major manufacturing defects.

The problems become more severe when the systems that we are dealing with are intended to perform data processing, operating on data supplied to the unit from an external source. Such systems are nowadays usually largely, if not entirely, digital in nature; at least three features of such systems give rise to testing problems.

a   For mechanical convenience, a large system is broken into sub-units mounted on separate printed circuit boards (pcb's). Each pcb needs to be tested separately but, in isolation, the pcb often has no identifiable function, so that the 'common-sense' approach outlined above is not possible.

b   The nature of the devices used in a digital system make it very unlikely that the output could ever be other than digital in nature. Inspection of the form of the output signals will, therefore, tell us very little about the functioning of the circuit. In other words, it is not enough that the output consists of 0s and 1s: the pattern of 0s and 1s has got to be such as to convey the right information.

c   Since the output depends on the data fed in, there is clearly a need to choose particular data with which to exercise the circuit. How this choice should be made is far from obvious. For example, if we have a calculator circuit which is required, among other things, to add two numbers, we may verify that it will correctly add two and two; is there any guarantee that it can also correctly add four and four? To verify correct operation for each possible pair of operands would be a daunting task; the choice of an adequate subset is scarcely less daunting.

The need for testing also arises in connection with the repair of faulty products, including both the results of unsatisfactory manufacture and parts that have failed in service in the field. Again, the wheelbarrow needs little in the way of formal procedure; the nature of the fault will itself usually define what is wrong and what is needed to put it right. Electronic circuits are more difficult; the diagnosis of the cause of failure has always been a skilled job, based usually on a process of tracing signal flow through the circuit with the aid of instruments such as voltmeters and oscilloscopes or, for digital circuits, logic probes and logic analysers.

The seriousness of the problems of testing electronic circuits has become apparent relatively recently, largely as a result of developments in micro-electronic technology. The two main effects of this technology are that component sizes have been dramatically reduced while the complexity of units at all levels (chip, pcb, system) has been dramatically increased.

The increased complexity makes an intuitive approach to testing ineffective; while the reduction of component size, reflected in the amount of circuitry that can be accommodated on a single chip, means that testing based on the use of measuring instruments to trace signal flow through the circuits cannot be performed. The magnitude of the changes in size and complexity can be illustrated by recalling that a typical minicomputer in 1970, including input, output and back-up memory, occupied perhaps two 19 in. racks each six feet high. This is about twelve times the volume of a 1985 personal computer, which, moreover, offers far more computing power and facilities than the mini had.

At the same time that the testing problem has been becoming more acute, manufacturers have been experiencing pressures that make testing more important. Product quality is being seen more and more as being vital to success; no matter how good the after-sales service may be, if faulty products are shipped in any quantity the effects on the Company can be both damaging and long-lasting. Product quality, however, is not enough on its own; the price must also be competitive. To keep prices down, automation must be applied to all stages of the production process; in particular we cannot afford the time required to execute a test procedure manually, neither can we afford to employ the skilled manpower needed to devise and administer such a test programme.

It is important to recognize that economic considerations are at the heart of all testing problems. Basically, testing costs money, and good quality testing (that is to say, testing that uncovers a large proportion of the faulty units that are produced) will cost more than poor testing. However, failing to test, or not testing adequately, also costs money in that a faulty product that escapes detection during the production stage will none the less eventually have to be put right, entailing the use of diagnosis and repair routines that cannot, by their very nature, be fully automated. The cost of not testing, therefore, can be very high: in addition to labour costs for carrying out the actual repair (skilled labour again), there are considerable administrative costs involved with any form of special treatment that takes place outside the standard production process. It is a matter of fine economic judgement to choose the quantity and quality of testing such that the cost of this testing is less than the consequential costs of not testing. The calculation is made more difficult because none of these costs can be known in advance; they depend on actual production failure rates. It has been said, truly but unhelpfully, that the best policy is to make the product correctly in the first place, in which case testing could be dispensed with! However, in the real world we have to recognize that no process can be perfect, so that testing, and, in particular, automatic testing, will be an essential part of production for the forseeable future.

1.2 THE PLACE OF TESTING IN MANUFACTURE

1.2.1 Integrated circuit fabrication

In modern systems, the smallest physical unit with which we have to be concerned is the integrated circuit (ic). In this book, we will confine our attention to digital ics.

On completion of the ic fabrication stage, we are left with a wafer containing many copies of the circuit arranged in a rectangular matrix. The next stage of the process is to slice the wafer into individual chips, each of which must then be encapsulated to form the familiar ic package. On the way, connections have to be wire-bonded between the chip and the ic pins. The obvious place of testing in this process might seem to be at the end, to verify that the completed chip functions correctly. However, two important points about fabrication technology influence the way that testing is incorporated into the process, and lead to the adoption of a different procedure:

a In the present state of the art (and, almost certainly, for some time to come yet) the proportion of non-working circuits emerging from the fabrication process is still large – typically, tens of per cent for a complex chip.

b The cost of bonding and encapsulation is a significant fraction of the production cost – perhaps up to a third.

As a result, it is worth while to conduct tests on the wafer before the chips are packaged individually. A simplified flow chart for a typical process is shown in Fig. 1.1, where each decision box represents a test.

Before considering whether or not the individual chips are working, it is

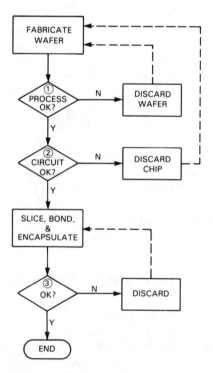

Fig. 1.1   Flow chart illustrating the ic manufacturing process.

usual practice to make a general assessment of whether the processing has been carried out satisfactorily. This is done by examining standard circuits known as **drop-ins**, a few copies of which (four or five perhaps) are included on every wafer. These drop-in circuits are chosen for their sensitivity to variations in various process paramaters; they are tested (test 1 in Fig. 1.1) by making electrical measurements of such properties as thresholds, noise margins and propagation delays. A wafer whose drop-ins display unsatisfactory performance will be discarded at this point, since the chance of finding working circuits on the wafer is too small to justify the cost of the next stages of manufacture, but information about the nature of the failure may be fed back and used to modify the process so as to increase the yield on subsequent production runs. This possible feedback is indicated in Fig. 1.1 by a dotted line.

Having established that the process is operating correctly, the next step is to transfer attention from the drop-ins to the actual circuits that are being manufactured. Because of the costs of bonding and encapsulation, a test must first be applied to the circuits while they are still part of the wafer (test 2 in Fig. 1.1). This test will normally be concerned with applying and observing logic signals, rather than the analogue measurements that were involved in testing the drop-ins. It is particularly significant here to observe that for this wafer test access to the chip is restricted. It may be thought that, since the interior of the circuit is exposed, all parts of it are accessible. Techniques for electronic probing have been developed, but they are difficult to apply in a fully automated system, and they have not found much favour outside research and development laboratories. In practice, therefore, access is by way of mechanical probes, and the only points of the circuit at which these can be applied without causing irreparable damage are the specially provided pads used subsequently for bonding to the ic pins. Additional pads to allow additional access for wafer testing could in principle be provided; but since they are (relatively) very large structures, the number of such extra pads that can be accommodated is always small and often zero. This restriction on access to the circuit has far-reaching effects on the methods that can be used to test the circuit.

Once the working circuits on the wafer have been identified, they are sliced, bonded, and encapsulated. These processes are also not infallible, so that a final test (test 3 in Fig. 1.1) is required before the product is finally shipped. This test again must be conducted with access limited to the input/output pins; failed chips can only be discarded since they cannot be repaired, but statistical or diagnostic information about the failures may be fed back to correct deficiencies in the bonding and encapsulation processes.

1.2.2 Assembly of pcbs

The next stage in the building of an electronic system is to put together a number of ics mounted on a pcb. A group of pcbs is then commonly mounted on a mother board to form the next level of system units. This progression is illustrated in Fig. 1.2, which serves to emphasize that each level represents a major increase in complexity compared with the one that precedes it. Once we have progressed beyond the chip level, a new very important feature comes into prominence – a faulty pcb is, in general, repairable. Since a large pcb may well have several hundred pounds' worth of components on it, and since the fault may reside in only one of these components or even in none of them (e.g. an open circuit in a track or a short circuit between two tracks), a simple decision that the board is or is not functioning correctly is not adequate: diagnostic information is required so that the fault can be remedied.

Techniques of diagnostic testing will be discussed in detail later; for the moment, we need only observe that the tracking of a fault to its exact location is a complex process, and, particularly, that the cost of diagnosis increases sharply with the level at which the testing is carried out. In terms of the boxes in Fig. 1.2, a rule of thumb that has gained widespread acceptance in the testing industry is that as we move from each level to the next, the cost of diagnostic testing increases by at least an order of magnitude. It is this basic economic consideration that accounts for the fact that, as with the ic, testing appears several times during the system manufacturing process. A flow diagram illustrating a typical process is shown in Fig. 1.3, which serves to high-

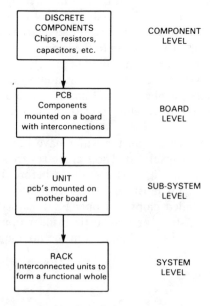

Fig. 1.2 Heirarchical view of a typical electronic system.

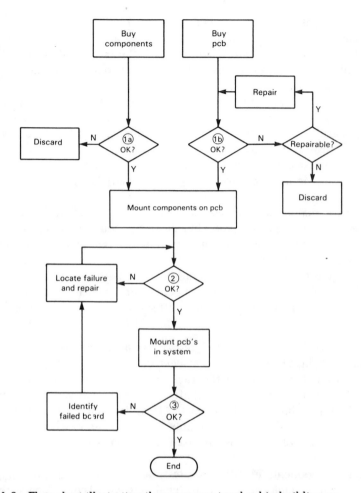

Fig. 1.3 Flow chart illustrating the processes involved in building an electronic system.

light some further important functions that testing is required to perform.

Three different levels of test can be distinguished in Fig. 1.3. At the first level are the two tests, 1a and 1b, on the component parts as they are received from the manufacturer. This is known as **goods inwards testing**; in the case of integrated circuits, test 1a serves the same purpose as, and could in principle be identical to, the final test of the ic manufacturer (test 3 in Fig. 1.1). This appears to be an unnecessary duplication, but is justified, at least for the more complex LSI or VLSI chips, by a number of practical considerations.

a Testing at the wafer stage will almost certainly be on the basis of testing 100% of the chips. At final test, however, it is common to employ sample testing; there is, therefore, a small but finite chance of a faulty chip escaping detection through not being tested.

b Even if a particular chip is one of those chosen for test, it can still turn out to be faulty when delivered. There are two main ways in which this can happen:

   i A chip that is good at test stage can suffer damage subsequently; mechanical handling can result in physical breakage of bonding leads or substrate, while electrical damage can be caused by electrostatic discharge.

   ii Human error can result in rejected chips being put into the 'accept' bin, or in chips (whether good or bad) being incorrectly type-marked.

c The ic manufacturer's test may well be insufficient to identify chips with marginal performance; but such chips could result in malfunction when incorporated in a complete system where greater demands are placed upon them.

d The most important consideration of all is that to identify a faulty component in isolation is very much easier, and hence cheaper, than to find the same faulty component after it has been incorporated into a system.

The other first-level test in Fig. 1.3, test 1b, is a **bare-board test** on a pcb. The rationale for this test is partly the straightforward economic one; it is quick and easy to find gross faults in the tracks of the board using a continuity test when there are no components to confuse the issue. Another reason for performing this preliminary check on the board is that certain faults (particularly short circuits to power supply lines) can carry risk of damage to components or even to the board itself if it is powered up with the faults uncorrected.

Faulty components identified by test 1a can only be discarded; pcb's rejected by test 1b may or may not be repairable. Good components are then mounted on good boards, which will then be tested as complete units (test 2 in Fig. 1.3). Two features of this test are particularly noteworthy:

a Even though it would, in many cases, be physically possible for the tester to gain access to any or all of the internal nodes of the circuit, the test program will often be written, as far as possible, with access restricted to the edge-connector. This again is an economic requirement; a program that demands more complicated access will require more expensive interface equipment and longer setting up times, resulting in a lower throughput. Access to points within the board is also both difficult and undesirable if, as is often the case, the pcb is finished with a protective coating intended to prevent deterioration due to atmospheric effects.

b As indicated in Fig. 1.3, a failure at level 2 will not normally result in the pcb being discarded; repair of the board is required, and this implies that a diagnostic test must be available. The diagnostic test is, in some cases, separate from the production test so that good boards can be processed quickly, with the generally slower diagnostic facilities being called in only when actually required for a faulty board.

When several pcbs are connected together to form a system or sub-system, the final test (test 3 in Fig. 1.3) takes on a form slightly different from that in level 2. The objective here is to apply an overall check on the function of the system; diagnosis is aimed only at identifying the faulty board. This board is then fed back to level 2 for detailed diagnosis and repair.

1.2.3 Maintenance and field service

Having conducted a series of comprehensive testing routines, the manufacturer can feel confident that he has shipped a fault-free system to his customer. However, his concern with the testing of his product does not end there. It has to be recognized that any system, however good, will, from time to time, fail in service, and will then need to have its failure diagnosed and repaired. In many cases, the servicing of the system remains the responsibility of the manufacturer; and although the customer will be paying for this servicing through a maintenance contract, the cost of the contract and the efficiency with which it is honoured will be important elements in the sales package and can have an important effect on the subsequent reputation of the manufacturer.

There are two special features of field service work that have a strong influence on the way the problem is dealt with:

a The customer is usually critically concerned with the down-time of his system. Even for a system that is just an ancillary facility (such as a telephone-answering system), the loss of that facility can be inconvenient and irritating; if the system is central to the customer's business (a computer that controls a milling machine for example), down-time represents loss of production and profits. Either way, the field service department will be under considerable pressure to get the system running again with a minimum of delay.

b Systems installed at a customer's site will usually be remote from the manufacturer's premises. The field-service engineer is therefore, in most cases, denied access to the range of testing aids that the production department has available, while at the same time being faced with the most difficult testing requirement – diagnosis at system level.

One way of dealing with the problem is to appoint permanent on-site maintenance staff and equip them with appropriate test equipment, so that, helped by gradually acquired intimate knowledge of the particular installed system and its habits and peculiarities, they can keep the system running with a minimum of down-time. This arrangement is certainly effective, but it is also expensive; an installation has to be very large or very important to justify it.

With the more usual arrangement, where field-service engineers have to deal with system failures as they arise, the problems of fault diagnosis and repair have to be faced. One technique that has traditionally been widely used is **board-swapping**, based on the idea that the field-service engineer, equipped with a complete spare set of boards, simply changes the boards in the system one by one until the fault goes away. The boards that have been removed, at least one of which is presumably faulty, are then returned for repair to a central depot; this may be the manufacturing site, but in any case, it will be equipped with, or have ready access to, the test equipment and test programs used in the original production process.

There are two important merits that can be claimed for the board-swapping approach to field servicing.

a The method is a straightforward one, so that, in most cases, the time taken to restore the system to operation will be minimal.

b Because the method is an unsubtle one, not involving detailed diagnosis, and requiring only an observation of the overall system operation, the field-service engineer does not need a high level of skill, and neither does he need very detailed knowledge of the system. This is a particularly important feature, since highly skilled technical staff are not only expensive but also in great demand and in short supply.

These merits must, however, be set against the very substantial costs incurred. Because it takes time for swapped boards to reach the repair depot, be diagnosed and repaired, and to get back to the field where they can be available for use again, there is a 'pipeline' full of boards the cost of which has to be borne by the manufacturer. Furthermore, there are consequential costs incurred in moving these boards from one place to another – postage, packing, insurance; even import and Customs duties may be payable if the boards cross international boundaries; and to keep track of it all will require administrative machinery. The only way of reducing these pipeline costs is to have board repair carried out on site; if this is to be done without substantially increasing the skill demands on the field-service engineer, he must be equipped with portable test equipment providing automatic diagnostic facilities. Thus the requirements of field servicing finally become translated into pressure first on the equipment manufacturers to provide suitable equipment, and then on the test department to provide compact but comprehensive test programs.

1.3 AUTOMATIC TEST EQUIPMENT

1.3.1 Typical architecture

The testing of electronic systems is based ultimately on the application of particular signal waveforms to the input terminals of the **unit under test** (UUT), and the observation of the resulting signals appearing at the output terminals. From measurements of this kind it is possible to derive parameters such as gain, resistance, risetime, frequency response, leakage current and so on. Such an arrangement can be automated by using signal sources and measurement devices whose operation can be controlled by sequences of switching signals, provided that the necessary measurement values are machine-readable. In practice, this is achieved by using digital instrumentation. **Automatic test equipment** (ATE) is based invariably on a digital computer system, often incorporating more than one processor; a general schematic is shown in Fig. 1.4. The UUT is connected through a test head fixture whose function is to supply power to the UUT and to provide communication channels between the signal sources and the UUT, and between the UUT and the measurement devices. The whole operation is under the control of the computer, which also has the task of collecting and evaluating the test results. The system is supplied with back-up memory (tape, disk, floppy, etc.) and the usual range of input and output devices (keyboard, VDU, printer, etc.) to allow operator interaction.

The structure illustrated in Fig. 1.4 is that of a parametric tester, which applies analogue stimuli, and takes measurements of analogue responses.

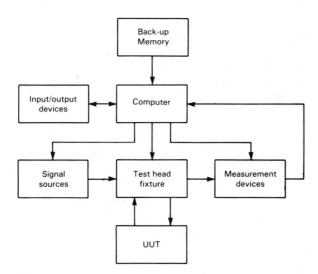

Fig. 1.4   Structure of an automated test system based on analogue measurements.

However, for digital circuits, the main testing task, which is performed by a logic tester, is the application of logic input signals and the observation of the resulting logic output signals (see section 2.1.2).

A tester architecture intended specifically for logic testing is shown in Fig. 1.5: it takes on a form slightly different from that of the parametric tester of Fig. 1.4. The most important difference is that since input stimuli are simply logic signals, the test program takes the form of a set of binary vectors that are applied to the UUT in succession; it is convenient to store these vectors in a local memory, often referred to as **memory behind the pins**. This allows much faster test application rates than would be possible if each test vector had to pass through the computer. For economic reasons the local memory will usually have a capacity substantially smaller than the length of a typical test program, and it will therefore have to be reloaded several times; this will be done under DMA from the back-up memory. An alternative source of test vectors is sometimes provided in the form of an **algorithmic pattern generator**, which is likely to be an ancillary system controlled by a separate microprocessor. It is particularly useful for testing highly repetitive structures (especially memories) where the test program consists essentially of the same basic pattern applied in turn to many different points in the circuit (see section 6.4). In such cases, it is economical of storage space to describe the required sequence mathematically in the form of an algorithm, and generate it at the time that it is required.

1.3.2 Pin electronics

An important feature in the structure of Fig. 1.5 is the set of **driver/sensor pins** that form the interface between the test program and the UUT. The electronics to service these pins has a crucial effect on the performance of the tester, and accounts for a substantial fraction of its cost. In the first place, we should observe that, because of the increasingly widespread use of bi-direc-

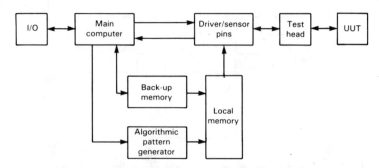

Fig. 1.5   Architecture of a typical logic tester.

tional data busses in modern electronic systems, it is essential that a test system should be able to treat any connection with the UUT as either an input or an output, and that, for some of the pins at least, it should be able to change the definition within the program under software control.

A typical driver/sensor unit for a logic tester takes the form shown in Fig. 1.6. It should be observed that, even although the test signals are expressed as logic values, the requirements must be translated into analogue voltages before they can be applied to the driver pins; equally, a sensor will receive an analogue voltage from the UUT, and must translate this into a logic value before it can be processed by the computer. The translation processes are undertaken by the blocks marked 'driver' and 'sensor', each of which is an operational amplifier circuit, fed with analogue reference voltages representing 0 and 1. These reference voltages are held on sample/hold circuits, having been established as part of the program set-up; notice that the analogue interpretation of the logic levels is, in general, different depending on whether we are driving or sensing. In the most flexible arrangements, each pin has its own set of definitions any of which can be changed during the execution of the test program. This flexibility, however, is clearly an expensive option; many testers are provided with cheaper but less flexible schemes, where pin values are defined in groups or as one of a few sets of values.

The working of the system of Fig. 1.6 is basically quite straightforward. If, at some particular point of the program the pin is being used as a driver, the driver/sensor input will enable the driver circuit. The logic value required on the pin is entered into latch 1 when the clock signal appears, and the driver circuitry then produces the appropriate analogue voltage (V_{DH} or V_{DL}) at the driver/sensor pin. The use of the latch ensures that inputs are applied at all the driver pins simultaneously. If the pin is to be used as a sensor, the driver/sensor input disables the driver circuitry, and the expected logic value is entered into latch 1. The analogue value appearing at the pin is compared,

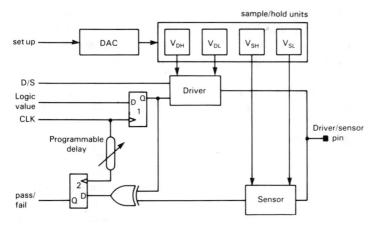

Fig. 1.6 Essential features of pin electronics.

in the sensor unit, with the reference values V_{SH} and V_{SL} so as to produce a logic value which is in turn compared in the XOR gate with the expected value from latch 1. The result of this comparison, which is interpreted as 0 = pass; 1 = fail, is clocked into latch 2 at a defined time after the inputs were applied – this programmable delay is necessary to allow the circuit to reach equilibrium, allowing for propagation delays.

1.3.3   Test head fixtures

In order for the tester to be able to apply test patterns to the UUT and to assess the resulting outputs, it is clearly necessary to provide physical connections between tester and UUT. The kind of test head fixture fitted to any particular tester will depend largely on the application for which the tester is intended; different types vary considerably in cost, in the kind of facility offered, in the time needed to set up the equipment for each new UUT, and in its suitability for fully automatic operation. Some important characteristics of fixtures in common use are summarized below.

   a A **wafer prober** consists of a set of microprobes arranged so as to make contact with the bonding pads round the periphery of a chip. It follows that this has to be customized for the particular chip, unless the manufacturer is prepared to work with a standard pad layout. In practice, demands for maximum utilization of chip area make it difficult to insist on a standard pad layout, so wafer probers invariably make some provision for using different probe geometries. The prober has an integral microscope, which is necessary to allow the operator to adjust the initial alignment between probes and wafer; a stepping mechanism then allows each of the chips on the wafer to be tested in turn automatically.

   b A **bed of nails** fixture consists of a matrix of spring-loaded pins, each of which makes contact with one of the circuit nodes on a pcb. A large pcb may have several thousand pins, so that the problem of making good contacts with all of them is a serious one; a vacuum chamber is usually fitted so as to provide a large and uniform pressure all over the board. A fixture of this kind is clearly very expensive, and, since it must be customized for a particular pcb, it will not be economic to use it except with large-volume production.

   c **Edge-connector** fixtures provide the most economical means of connecting a pcb to a tester, because the fixture does not have to be customized to the pcb. However, if access to the circuit is confined to those nodes connected directly to the edge-connector (the primary inputs and outputs), it may not be possible to generate an adequate test program. (This point and its implications will be discussed at length in later chapters.) Additional access in such cases can be provided by attaching clips or probes to selected internal nodes of the circuit, and connecting them to unused driver/sensor outputs. Such additions to the interface are used sparingly because:

i   it implies a degree of customization of the fixture;
ii it increases the time needed to set up the tester for each new board;
iii the set-up process becomes more error-prone.

One additional facility that is almost invariably provided with edge-connector fixtures, however, is a **guided probe** which is normally used in the diagnosis phase of testing; the tester is able to resolve diagnostic ambiguities by collecting data from selected internal nodes, and obtains this data by issuing instructions to the operator who positions the guided probe accordingly.

1.3.4   Types of ATE

The range of applications for which a tester can be used is strongly influenced by the kind of test head fixture that is fitted. It would, in principle, be possible, by fitting the tester with a range of alternative fixtures, to make a general-purpose ATE capable of applying any of the different tests that appear in the schematics shown in Fig. 1.1. and 1.3. Such a system, however, would be extremely expensive; it would also be a very uneconomical use of capital, because many of the facilities would necessarily be idle at any one time. By making test systems directed at particular applications, it is possible to minimize the cost while optimizing the facilities provided in terms of:

a   the test head fixture;
b   the number of driver/sensor pins;
c   the provision of automatic handling equipment;
d   the measurement system (hardware and software).

The classification of ATE into distinct types is not always easy, and the terminology is not always consistent, but it may be helpful to introduce some of the terms that will be found in technical and sales literature.

One fairly clear distinction has already been drawn between **parametric testers,** which measure analogue quantities such as leakage current and output voltage, and **logic testers,** which measure logic states. Our concern in this book will be entirely with logic testers.

The first application area for which dedicated test equipment is normally used is that of wafer testing. This requires the use of a wafer prober, and can take either parametric or logic form. The test head is expensive, but on the other hand the number of driver/sensor pins required is relatively small, so that the cost of pin electronics (the most expensive part of the electronics) is kept down. No automatic handling equipment is involved; it is not possible to load the tester automatically since each individual wafer has to be lined up by hand, although the subsequent testing of each chip on the wafer is then automatic. An ink-jet attachment is usually fitted on the test head so that chips that fail the test can be marked for subsequent disposal.

A very large class of specialist ATE is formed by **component testers**. These come in many different forms and the facilities provided are determined by the type of component for which the tester is intended. Component testers are used for the final test made by the component manufacturer (test 3 in Fig. 1.1.) and for the goods-inwards test made by the system manufacturers (test 1a in Fig. 1.3). For all such testing an automatic handling system is needed, so that each sample is taken from a bin, loaded into the test head, and put into categorized output bins without any operator intervention apart from keeping the input bin topped up and removing the tested components. It should be noticed in passing that, small though the need for operator intervention is, it is enough to allow human error to influence the process. The problems were mentioned in section 1.2.2, and are simply due to misbinning: failed components can be inadvertently put into the 'good' bin, or components of one type can be put into the bin corresponding to another type, resulting in the component being incorrectly labelled. Both of these effects are found in small but significant quantities; these effects among others help to justify the double check provided by goods-inwards testing.

The complexity of a component tester depends very much on the kind of component that is to be tested, varying from the simplicity of machines to test discrete devices (such as capacitors, resistors, and diodes) to the much more elaborate capabilities required of a machine intended to test a VLSI chip. Memory chips, in particular, have very special testing requirements (these will be discussed more fully in Chapter 6) and are usually dealt with by a specialist tester, often one capable of testing a number of identical chips simultaneously.

There are two quite different approaches to the testing of pcbs. **In circuit testing** is based on the use of a bed-of-nails fixture, which provides access to the input and output pins of each individual component. This permits each component to be tested in isolation, provided arrangements can be made to overcome the influence of surrounding components – a process known as **guarding**. This is perhaps the commonest method in use today for testing large pcbs. It has the merit that test pattern generation is simple, basically just a matter of a collection of standard library routines. To set against this, however, there are some major disadvantages:

a the test head fixture is very expensive;
b a separate fixture is needed for every pcb;
c the number of pins, and hence pin electronics, is very large, so that the tester is very expensive.
d testing that all the components are working does not guarantee that the system as a whole is fault-free.

The widespread use of in-circuit testers despite all these objections is an indication of just how serious are the problems of test-pattern generation.

One special case of the in-circuit type of tester applied to a slightly different situation is the bare-board tester. This is used to apply test 1b in Fig. 3 – essentially the goods-inwards test for pcbs, as discussed in section 1.2.2. The bare-board tester uses a bed-of-nails fixture to access all the tracks on the board, after which detection of breaks in the tracks and short circuits between tracks is a simple matter, making few demands on either test facilities or on test program generation effort. However, the cost of the test head fixture will be high and this will imply that bare board testers will be used only in high volume production applications, and in such cases will normally have automatic handling equipment to increase the throughput.

1.4 THE TEST ENGINEER'S JOB

1.4.1 The economics of testing

Throughout this chapter, the costs of hardware, software, and human resources have been repeatedly mentioned. Economic considerations are without doubt central to the whole business of testing. The manufacturer seeks to provide his customer with a product that meets its specification; the amount and quality of testing to be employed, and the stages in the production process at which the testing is to be applied, are chosen with a view to minimizing the overall cost of achieving this end. It has already been remarked that in an ideal world in which production processes were perfect, testing would be unnecessary; each time we introduce a test, it can be regarded as an insurance premium that is paid with a view to reducing the penalty attached to the occurrence of a manufacturing fault. As with any other form of insurance, the economic justification depends on the relationship between, on the one hand, the premium involved, and, on the other hand, the difference between the cost of correcting faults with and without the insurance.

Deciding on the optimum testing policy is complex, since there are many possible variations both in the types of testing to be employed and in the ways in which the tests are fitted into the production line; the decisions are complicated particularly by the fact that all the cost figures have to be based on estimates. The costs and benefits arise in several ways:

a The costs of providing any particular type of testing are divided between
   i the capital cost of the equipment and of producing test programs;
   ii the running costs (operating and maintenance). The cost per unit will depend on the throughput and the life of the equipment. This calculation will be further complicated by the need to consider the likely market for the unit, since any product-specific components of

cost (such as test programs and bed-of-nails fixtures) will have to be written off when the particular unit goes out of production.

b The returns on the investment in testing depend on the failure rates at each stage in the process and on the associated cost of recovery from failure. At the time at which decisions have to be taken, these figures are not available, so that estimates based on past experience have to be used; the accuracy of these predictions will have a profound effect on the overall calculations.

This, then, is one aspect of the test engineer's job; the decision to use a particular form of test has enormous implications in terms of capital investment. Further cost implications follow from the choice of test strategy. Basically, this is the choice of what to do with a failed unit; among the options are

a discard the unit;

b repair the unit and return it to repeat the test that it had previously failed;

c repair the unit and send it direct to the next stage of the production process.

The objective is to minimize the costs of testing; it is worth pointing out that these costs can contribute very significantly (commonly 10–20%, and in some cases 50% or more) to the overall cost of a component.

1.4.2 Test pattern generation

Important as the choice of test strategy undoubtedly is, further discussion of the topic lies outside the scope of this book. We will be concerned principally with the other major activity of the test engineer – preparing the test programs that enable the particular units to be exercised on the particular ATE that is in use. Ideally, these programs would be prepared automatically by a computer; but this ideal has yet to be realized except in special cases. Manual methods will therefore be discussed in detail; such methods are not only useful in themselves but also provide the basis for such automatic methods as exist.

1.5 THE IMPACT OF TESTING ON DESIGN

Once the problems involved in producing test programs for circuits and systems have been fully appreciated, it soon becomes apparent that the circuit designer has it in his power to make a major contribution to the ease with which his circuit can be tested by adapting his designs, or by adopting appropriate design philosophies. This aspect of the subject is becoming increasingly recognized by the circuit-design community, and nearly all major manufacturers now claim to impose more or less constraints on the design as concessions to testability. The principles on which such design constraints are based will be discussed at some length in later chapters of the book; for the moment, it is sufficient to observe that, starting from the position of testing as an activity to be pursued after the product has been made (and not to be thought about until then), the demands of technology have led first to the incorporation of test procedures throughout the manufacturing cycle, and finally to the design itself being governed, or at least constrained, by the testing requirements.

SUMMARY
Chapter 1

Advances in microelectronic technology over the past few decades have brought about dramatic changes in the complexity and size of electronic circuits. These developments are continuing still, and seem likely to do so for some time yet, and are making it increasingly difficult to define adequate test procedures. At the same time, testing is occupying an increasingly important place both in manufacture, where it is needed to guarantee the quality of the product, and in maintenance and field service, where it is needed to provide diagnostic capability.

Testing is an expensive process, but the consequences of inadequate testing within the manufacturing process can be even more expensive in the long run. The choice of a testing strategy is a matter of economic judgement based on the costs of test equipment, the predicted failure rates, and the consequential costs of failing to find faults at any particular stage in production.

Automatic test equipment generally consists of a sophisticated computer-based system that manages the test data and analyses the responses, together with specialized (and generally expensive) interface circuitry. Different types of ATE, intended for use at different points in the manufacturing process, are distinguished chiefly by the form of the test head fixture.

EXERCISES
Chapter 1

E1.1 The manufacturer of a complex electronic system may decide to use Goods-Inwards Testing for some or all of the component parts. Discuss the factors that need to be considered in coming to this decision.

EXERCISES
Chapter 1
continued

E1.2 In a block diagram describing the architecture of an automatic test system, features often found include:

a memory behind the pins;

b guided probe.

Explain what these features involve, and why they are needed.

E1.3 Explain the principle of board-swapping as a field-servicing strategy. Discuss the advantages and disadvantages of the technique.

E1.4 Describe the various forms of test head fixture that can be fitted on ATE. Compare the capabilities of these fixtures.

E1.5 What are the important requirements of the pin electronics in an ATE? Describe a typical system and explain how it works.

2

TEST-PATTERN GENERATION

2.1 BASIC TERMINOLOGY

2.1.1 Circuit description

The circuit represented diagrammatically in Fig. 2.1 serves to illustrate some of the common structural features that are significant from the testing point of view. Certain nodes of the circuit are described as **primary inputs** (PI) or **primary outputs** (PO); in an ic these are the pins, while in a pcb they are the edge-connectors fingers. In either case, they are the points at which interconnections are made between the circuit and the rest of the system. In Fig. 2.1 these nodes are shown as A_1–A_n and Z_1–Z_m. Other nodes of the circuit, such

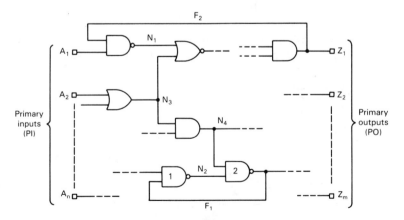

Fig. 2.1   General combinational logic circuit.

as nodes N_1–N_4, which are not brought out to input/output terminals, are called **internal nodes**; it is often inconvenient and sometimes impossible to access internal nodes while testing.

Figure 2.1 also shows two feedback connections, F_1 and F_2. A distinction is sometimes drawn between **local feedback**, which is applied round a small sub-section of the circuiut, and **global feedback**, which is applied round the whole circuit or a major part of it. The distinction is not always clear cut, but it roughly corresponds to whether or not the effect of the feedback is obvious. F_1, for example, is a local feedback path, which has the effect of turning gates 1 and 2 into a bistable latch. F_2 is a global feedback path, which will also introduce time-dependent behaviour into the circuit, but the precise nature of that time-dependent behaviour can be determined only by carrying out a detailed analysis of the circuit.

Another significant feature of the topology of a circuit is exemplified in Fig. 2.1 by nodes N_3 and N_4. In almost all practical circuits there will exist at least one node at which a single signal is routed to the inputs of two or more gates. This feature is described as **fan-out**, and it brings with it some additional complications from the testing point of view.

The circuit shown in Fig. 2.2 contains two examples of fan-out; the input node B and the output of gate 3 each fan-out to provide inputs to two other gates. These two examples of fan-out exhibit slightly different properties because of another topological feature. Wherever we have fan-out there will be two or more pathways between the fan-out node and the primary outputs. These pathways can either remain separate, or can subsequently come together again as inputs to a single gate. In the circuit of Fig. 2.2, for example, the output of gate 3 has one pathway through gates 4 and 6 to output Y, and another through gate 5 to output Z; whereas the two pathways leading from the fan-out at node B, one through gates 1 and 6 and the other through gates 2, 4 and 6, eventually recombine at the output of gate 6. This is an example of **reconvergence**; the distinction between reconvergent and non-reconvergent fan-out will be important later on when considering fault-equivalence.

The process of testing the circuit will consist of applying successive sets of values to the primary inputs, and of observing the resulting values appearing

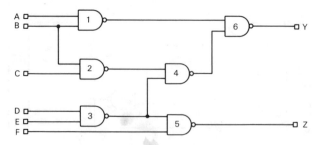

Fig. 2.2 Circuit illustrating fan-out and reconvergence.

at the primary outputs. In order to assess the result of a test, it will be necessary to know the output values that would be obtained from a correctly functioning circuit; these are called the **fault-free outputs**. Each individual test, consisting of a set of input values together with the corresponding set of fault-free output values, is known as a **test pattern** or **test vector**. A complete sequence of test patterns is called a **test set**.

2.1.2 Types of test

There are two broad classes of measurement that will enable the behaviour of a circuit to be described (see section 1.3.1).

a **Parametric testing** seeks to measure analogue quantities such as voltage, current, time, etc. Such measurements are, of course, essential to the testing of analogue circuits; they are also significant in digital circuits, where incipient failure could often be deduced from correct but marginal behaviour. Degraded logic levels or excessive propagation delays, for example, can reduce noise immunity or introduce glitches and other timing problems.

b **Logic testing** is concerned with digital measurements; this implies a thresholding operation (as discussed in section 1.3.2) to convert analogue voltages into logic values.

In this book we will be concerned essentially with logic testing, although the importance of parametric testing should not be underestimated. It is not uncommon to find a pcb that contains a mixture of analogue and digital components – a unit such as a disc controller is a typical example – and such circuits present particularly difficult problems to the test engineer. These problems will not be considered further in this book, largely because they are essentially unsolved – such circuits tend to be at best tested using intuitively derived test procedures which are acknowledged to be inadequate, and at worst not tested at all.

2.1.3 Functional test pattern generation

The central problem of testing is the derivation of an adequate test set for any particular circuit. This process is described as **test pattern generation** (TPG); ideally, it would be performed entirely automatically by a computer as part of a CAD facility, but this ideal has been fully realized only for a restricted class of circuits, and then only at the expense of program run-times of many hours on a large mainframe computer even for a circuit of quite moderate complexity. The algorithms employed by computers to provide automatic TPG (ATPG) are based largely on a formalization of a manual approach, and for this reason alone it is worth devoting some time to a consideration of manual methods.

The obvious approach to TPG is to exercise the circuit so as to check that it performs its intended function. A program developed in this way is known as a **functional test program**. A functional test of a three-input NOR gate, for example, could take the form of eight test patterns corresponding to the eight lines of the truth table that describes its operation. The concept of a functional approach has an appealing simplicity, but applying this concept to a practical circuit is not as easy as it appears.

Before starting to design any system, the first task facing the designer is that of preparing an accurate and comprehensive specification of the required function. This task has long been recognized as not only one of the most important in the design process, but also as one of the most difficult. For the test engineer, faced very often with nothing more than a circuit diagram, the task of identifying the function is even more difficult. There are problems with all types of circuit:

a For a simple combinational circuit, the use of the truth table (as suggested above) appears attractive. By covering all possible input patterns we are applying an exhaustive test set, which can be expected to check for all possible departures from fault-free behaviour. There are, however, two problems associated with an exhaustive test of a combinational circuit.

i For a circuit of any complexity the truth table will not be easy to derive; and it will almost certainly not be supplied to the test engineer.

ii A large and well-populated pcb may well have several tens of inputs. Quite apart from the difficulty of deriving the truth table, an exhaustive test is not practical. For n inputs the truth table contains 2^n lines; if we work through the table at 1 test/μs, this would take:

for $n = 30$, 18 min;
for $n = 40$, 13 days;
for $n = 50$, 36 years.

b For a sequential circuit the concept of an exhaustive functional test is even more difficult to apply. The problems of sequential circuits will be considered in detail in Chapter 5.

c LSI and VLSI components may appear in some cases to have an easily defined function, but rigorous definitions are not in fact easy. A microprocessor, for example, may be defined as a device that executes each of the instructions in its instruction set. An exhaustive test of the instruction set, however, would involve each instruction with all possible operands or combinations of operands; this will clearly suffer from the infinite time problem as with large combinational circuits.

Despite its limitations, some elements of a functional approach to TPG are often used for dealing with complex units, but an exhaustive test set is never attempted except for very small circuits.

2.1.4 Structural TPG and fault-models

Functional TPG seeks to find out if the circuit is working; the alternative is to seek to find out if the circuit is faulty. Programs developed on this basis are known as **structural test programs**, because they attempt to detect specific structural failures.

In structural TPG, attention is ultimately directed towards **defects** in circuits; these are the actual physical failures that result in functional failure.

There are many different kinds of defect arising from many different possible imperfections at various stages in the life of the product. If we are testing a pcb, for example, we can identify at least three different classes of defect:

a **Manufacturing defects in the ics**. Fabrication defects will have been largely caught at the wafer test stage; most of the ic defects that appear at the pcb test stage are basically mechanical. Imperfect bonding, for example, can result in contacts breaking while the chips is being handled; poor encapsulation, leaving air trapped inside the package, can lead to corrosion of the metallization.

b **Production defects on pcbs**. Defects in the manufacture of the pcb itself (such as broken or incorrectly routed tracks, or missing/extra connections at plated through holes) will most easily be detected in a bareboard test. The common production defects on the populated board arise in:

　i assembly: use of the wrong component, or inserting a component the wrong way round, or failing to align pins with holes, so that pins are folded under the chip;

　ii soldering: mainly dry joints resulting in no contact, or solder splashes connecting two tracks together.

c **Operational stress**. In the case of a pcb returned for repair after having been in the field, defects can be due to normal mortality of components (the result, perhaps, of thermal cycling or of continuous power dissipation in use), or components can be destroyed by electrostatic shock, or dirt can provide short-circuit paths.

The aim of a structurally-based test program is to ensure that a board containing any of these defects will fail at least one of the tests. Before considering how a test program can be constructed, it is first important to distinguish between a physical defect, such as those described above, and a **fault**, which is the electrical effect of that defect. An electrical test, consisting of applying electrical signals at inputs and observing electrical responses at outputs, cannot directly observe a defect, so that attention must be directed towards faults rather than defects.

The relationship between defects and faults is described by a **fault-model**. It is to be expected that any particular fault could result from any one of several different defects, so that a general-purpose fault-model could concentrate on providing a list of possible faults without necessarily relating the faults to particular defects. This is the principle on which the **single-stuck-fault model** operates; this model is widely used in the testing industry as the main basis for TPG.

The single-stuck-fault model consists essentially of two assumptions:

1 the fault directly affects only a single node in the circuit;
2 the effect of any defect is to make one particular node remain stuck at either 0 or 1.

If these assumptions were valid, it would follow that any test program that detected any single stuck node would cover all possible defects. In practice, both assumptions are open to question:

a Multiple faults are clearly possible, or even probable, since an original defect (a short-circuited diode, say) could well cause consequential damage to another component. There are two justifications for maintaining the single-fault assumption.

   i It seems intuitively reasonable to suppose that a circuit with two faults will still fail the test program. There has been some discussion in the literature of the possibility of the effect of a second fault cancelling the effect of the first fault – a process known as **fault-masking** – but very few test engineers believe that this is a significant problem in practice.

   ii The number of possible multiple-fault conditions in a circuit rises exponentially with the number of nodes, so that it is not feasible to consider all the possibilities.

b While many real defects are modelled adequately by the stuck-node assumption (e.g. short circuits to power rail or ground either inside or outside the chip; open circuits at the inputs to TTL chips, which usually have the same effect as logic 1; etc.) it is not difficult to find examples of defects that do not produce stuck nodes. The use of the stuck-node assumption is justified empirically; test sets produced on this basis appear to provide adequate protection in practice.

Despite the doubts about the validity of its underlying assumptions, the single-stuck-fault model is universally used in the testing industry as the basis for structural TPG, and as a standard of comparison when assessing the quality of a test set. It may be supplemented by some additional faults intended to cover particular defects that might otherwise escape detection; additional fault-models to cover these situations will be introduced later.

2.2 PREPARING A TEST PROGRAM

2.2.1 The strategy of structural testing

The basic procedure for structurally-based TPG (otherwise known as **fault-oriented TPG**) is represented in the flow-chart of Fig. 2.3. Each test in the set will be intended to cover one out of a pre-defined list of possible faults. The first requirement, therefore, must be to prepare the fault-list on which the test programme will be based. For the moment, we will confine ourselves to the use of the single-stuck-fault model, but the technique of TPG is essentially the same whatever fault-model is used.

Although each test may be written with the intention of covering one specific fault, it will almost invariably turn out in practice that the test will also cover other faults on the list. As a result, the number of tests needed in the final test programme will usually be considerably smaller than the number of faults in the fault-list. In order to avoid unnecessary duplication, therefore, each test should be checked as soon as it has been generated to establish its fault-cover and all faults that are covered by the test will then be removed from the fault-list.

Finally, the faults remaining on the fault-list will be inspected to determine whether or not the specified fault-cover target has been met. With a complex circuit, economics dictate that a limit must be placed on the effort that can be applied to TPG. For the simple circuits considered in this chapter we will seek

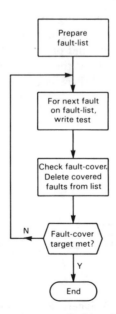

Fig. 2.3   Flow-chart illustrating structurally-based TPG.

100% coverage of the faults on the fault-list, but this would not normally be the target in a circuit of practical size.

2.2.2   The sensitive path concept

The central problem in TPG is ensuring that a fault appearing at the input of a circuit element produces an effect at the output of the element. The problem is illustrated in Fig. 2.4(a), which shows an element with four inputs, A, B, C and D, and one output Z. When considering the use of the element in a circuit, the function would be expressed by an equation of the form

$$Z = f(A,B,C,D)$$

For testing purposes, however, it is convenient to view the element in a different way. The point at issue is the relationship between the output, Z, and one of the inputs, A for example. This relationship can be manipulated by choosing particular values for B, C and D. Two particular forms of relationship are of interest:

a making Z dependent only on A;
b making Z independent of A.

In Fig. 2.4(b), for example, we have

$$Z = \overline{A} \text{ if } B = C = 1$$
$$Z = 1 \text{ if } B \text{ or } C \text{ (or both)} = 0$$

In the first case, B and C are acting jointly as 'enable' inputs, and they establish a **sensitive path** from A to Z. In the second case, we have a choice of ways of 'disabling' the element and providing a fixed value at the output. These forms of manipulation lie at the heart of all TPG methods.

It is interesting to note that there is one form of gate that cannot be disabled. The exclusive-OR gate, as shown in Fig. 2.4(c), always provides a sensitive path, since

$$Y = D \text{ if } E = 0$$
$$Y = \overline{D} \text{ if } E = 1$$

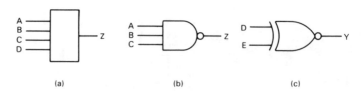

(a)   (b)   (c)

Fig. 2.4   Path sensitization. (a) General logic element; (b) three-input NAND gate; (c) XOR gate.

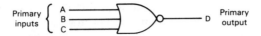

Fig. 2.5 Three-input NOR gate.

2.2.3 Testing an isolated gate

As a first illustration of the TPG procedure, we will consider the three-input NOR gate shown in Fig. 2.5, and work through the steps described by the flow-chart of Fig. 2.3, assuming that A, B and C are primary inputs and D is a primary output.

    a Preparing the fault-list. As we are confining ourselves to the single-stuck-fault model, this consists simply of each node stuck at each value. For convenience we can abbreviate node X stuck at 0 or 1, as X s-a-0 or X s-a-1; alternatively they can be represented as $X/0$ or $X/1$. Hence the fault-list for the NOR gate is

$A/0$ $A/1$ $B/0$ $B/1$ $C/0$ $C/1$ $D/0$ $D/1$

    b Writing a test. There are two distinct requirements that have to be satisfied in writing any test.

    i   If we are to test for $A/0$, we must establish a fault-free condition $A = 1$, otherwise there would be no difference between faulty and fault-free circuits.

    ii   If the presence of the fault is to be detectable at a primary output then the change to an incorrect logic value at the faulty node (the **fault-site**) must produce a change at a primary output; in other words, a sensitive path must be established between the fault-site and an output.

In the case of the NOR gate, if either B or C were at 1 then the output would be 0 irrespective of the value of A; the fault in this case would not be observable at the output, and would be said to be blocked. By making $B = C = 0$ we enable the gate, and the fault is then transmitted; that is to say, the value of D is determined entirely by the value of A.

    The test for $A/0$ is therefore to set $A = 1$, $B = 0$, $C = 0$; with this set of inputs the fault-free output is $D = 0$. It should be noted that the fault-free output value is an essential part of the specification; without it we have no means of knowing whether the circuit has passed or failed. Hence the test for $A/0$ can be written 100/0, where the value shown after the slash is the fault-free output. An alternative way of recording the test is $A.\overline{B}.\overline{C}/\overline{D}$.

    c Checking the fault-cover. A fault will be covered by a particular test if the change represented by that fault would be transmitted to a primary output. The cover can be assessed by:

    i   establishing fault-free values throughout the circuit;

ii for each node in turn, determining whether a change in its fault-free value would produce a change at a primary output. This is best done by starting from the outputs and working back towards the inputs.

In this case we can deduce:

i $D/1$ is covered. (Notice that every combination of inputs constitutes a test for each output stuck at the inverse of its fault-free value.)

ii $A/0$ is covered. (This was the fault for which the test was written.)

iii Since $A = 1$, transmission from B or C to the output is blocked so that $B/1$ and $C/1$ are not covered by this test.

This completes the first cycle of the TPG process. Having removed $D/1$ and $A/0$ from the fault-list, we return to repeat the whole process with the next fault; in this case $A/1$. Using exactly the same techniques as before we obtain:

a $A = 0$ to establish the fault-sensitive condition;
b $B = C = 0$ to enable the gate;
c Hence test is $000/1$;
d Fault cover: $D/0$, $A/1$, $B/1$, $C/1$

The only faults now left on the list are $B/0$ and $C/0$: these are covered by $010/0$ and $001/0$ respectively.

2.2.4 Testing a combinational circuit

The testing of an individual gate, where we have direct access to all the inputs and outputs, does not present any great difficulty. A slightly more realistic problem is represented by the circuit shown in Fig. 2.6, where the significant new feature is the existence of internal nodes (E and F) which are neither primary inputs nor primary outputs. This feature does not alter the basic method of tackling the problem, but it does make it a little more difficult to apply, in that, in the absence of direct control of any particular node, it becomes necessary to establish a required value indirectly from the primary inputs.

Consider, for example, a test for $A/0$ in the circuit of Fig. 2.6. The steps in this process would be

a $A = 1$ to establish the fault-sensitive condition;
b $B = 1$ to enable gate 1 and transmit the fault to node E;
c To transmit the fault through gate 3, so as to complete the sensitive path

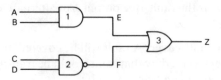

Fig. 2.6 Circuit with internal nodes.

from the fault-site to the output, we require $F = 0$. This is an example of a requirement to establish a fixed value at an internal node of a circuit. To satisfy this requirement we have to work back towards the input end of the circuit so as to determine suitable values at the primary inputs. This need to alternate forward and backward tracing through the circuit is characteristic of TPG activity.

In the circuit of Fig. 2.6, the fixed value $F = 0$ is established from the primary inputs with $C = D = 1$.

Hence the test is 1111/1.

So far each fault we have investigated has given rise to one and only one test. This does not always happen, however; in practice, especially in circuits with many inputs, many of the faults will be covered by several tests. Consider, for example $F/1$ in the circuit of Fig. 2.6. This requires $F = 0$, which is established by $C = D = 1$. To enable gate 3, we also require $E = 0$. Since E is the output of an AND gate, it can be held at 0 by supplying 0 to either or both of its inputs. Hence there are three tests that cover $F/1$: 0011/0, 0111/0, 1011/0.

2.3 FAN-OUT AND RECONVERGENCE

One complication that is sometimes introduced by reconvergence is illustrated by the circuit shown in Fig. 2.7. Suppose we apply the test pattern 111/1, and consider the effect of the fault $B/0$. The changes introduced by this fault are shown in Fig. 2.7, where the symbol $1 \rightarrow 0$ indicates a fault free value of 1 that becomes 0 under the particular fault condition under consideration. It will be seen that, because of the fan-out, the fault will affect the inputs of both gates 1 and 2, and, because of the values assigned to A and C, the fault is transmitted through both gates, and hence affects both inputs of gate 3, which is where the reconvergence takes place. Since gate 3 is a NAND gate, a change of inputs from 00 to 11 will produce a change of output from 1 to 0, and hence the fault is transmitted through gate 3. In this case, although neither input change alone would have produced an observable change at the output (since 01 and 10 both give the same output as 00), the two acting together result in fault transmission. (This mechanism is known as **dual-path sensitization**.)

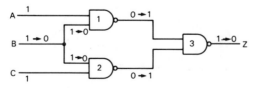

Fig. 2.7 Positive reconvergence through dual-path sensitization.

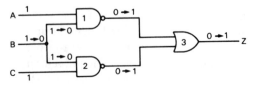

Fig. 2.8   Positive reconvergence where each path is separately sensitive.

If gate 3 had been an OR (or NOR) gate, as shown in Fig. 2.8, the situation would have been slightly different, in that each of the two input changes separately would have been transmitted to the output, but again, as in the circuit of Fig. 2.7, the two acting together would also produce a change at the output. In both these cases, therefore, the effects produced by the two input changes at the gate at which the reconvergence takes place reinforce each other; the situation is described as **positive reconvergence**.

A different situation is illustrated by the circuit of Fig. 2.9. If in this circuit the test 100/1 is applied and we consider the fault B/1, then we find that, at the input to G3, where the reconvergence takes place, the overall change is from 10 to 01, and so, since the output of a NAND gate in response to each of these input patterns is the same, the fault is not transmitted through G3. The effect, in which the fault transmission along one path is counteracted by the fault transmission along another path, is described as **negative reconvergence**. It should be noticed that the topology of each of the three circuits of Figs. 2.7, 2.8, and 2.9 is identical. Hence, although the existence of reconvergence in a circuit is easily detected by looking at the topology, topology alone is not sufficient to determine whether the reconvergence is positive or negative.

The implications of reconvergence on the testing process depends on the particular circuit being considered, and on the particular fault in that circuit. In the circuit of Fig. 2.8, the two pathways produce the same effect with respect to the fault B/0 in the test 111/0; there is no need in this case to take any special notice of the reconvergence. In the example of Fig. 2.7, some care is needed in generating the test; it would be easy to overlook the combined effect of the two pathways and observe only that each one on its own is blocked. Similarly with the circuit of Fig. 2.9, the checking of fault-cover has to be re-evaluated when the process gets back as far as a fan-out point. In all cases, the difficulties are not severe for a human doing the necessary

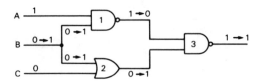

Fig. 2.9   Negative reconvergence.

operations manually, but they could well reveal themselves within a computer-based algorithm. We will return to this point in section 3.5.

2.4 UNDETECTABLE FAULTS

While generating a test for a particular fault, it will sometimes turn out that the procedure leads to a conflict. The problem may be illustrated by the circuit shown in Fig. 2.10, in which we might seek a test for $D/0$. Adopting the standard procedure yields the following:

a To detect $D/0$ we require $D = 1$. The only way of establishing this condition is to make $A = B = 1$.
b To propagate the fault through G5 requires $G = 0$.
c To obtain $G = 0$, the inputs of G4 must both be high, so that we need to produce fixed values of $E = F = 1$.
c The inputs to G2 are A and B, which are both high, so that the value of E is already determined: $E = 0$.

These conflicting requirements around G2 have been derived without having made any unforced choices, so that the problem cannot be resolved: this fault is undetectable.

The existence of an undetectable fault at the output of a particular gate must imply one of two limitations imposed on the gate by the structure of the circuit. Either:

a the inputs to the gate cannot be set to one or more of its 2^n possible states; or
b whenever the gate is set to the state of interest, other pathways ensure that the primary output values are independent of the gate output.

Whichever is the cause of the problem, it is almost invariably brought about by having redundancy in the circuit. Closer examination of the circuit of Fig. 2.10, for example, reveals that the overall function can be expressed as

$$Z = A + B + \overline{C}$$

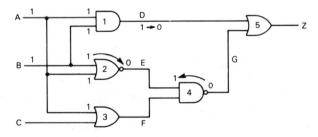

Fig. 2.10   Circuit with an undetectable fault.

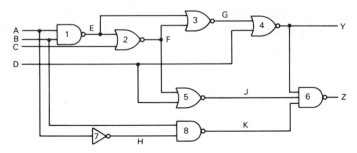

Fig. 2.11 Redundant circuit.

and that the output of G4 is also given by

$$G = A + B + \overline{C}$$

Hence, G1 and G5 are both superfluous; if they are removed, the undetectable fault will disappear.

In this particular example, specific redundant hardware is identifiable.
This, however, is not always the case. An attempt to find a test for $H/1$ in the
circuit of Fig. 2.11, for example, leads to a similar conflict as can be
demonstrated by the following analysis.

a Establish the fault conditions with $H = 0$. This requires $A = 1$.
b To transmit the fault through G8 requires $B = 1$. Hence $E = 0$.
c To transmit the fault through G6 requires $Y = J = 1$. To give $Y = 1$
requires $G = D = 0$; to give $J = 1$ requires $F = D = 0$.
d If now we look at G3 we find that its output G is required to be low, and
at the same time, both its inputs, E and F, are required to be low; hence
the conditions cannot be satisfied, and we deduce that the fault is
undetectable.

This circuit is a genuine example that appeared as part of a printed circuit
board for which a test program was being written, and it shows that, even
without any obvious hardware redundancy, testing problems can be brought
about simply by a failure to minimize logic. Analysis of the circuit of
Fig. 2.11 shows that its function is represented by

$$Y = \overline{D}(\overline{A} + \overline{B} + \overline{C})$$
$$Z = B + D$$

These functions can be obtained with the use of just three gates, as shown in
Fig. 2.12, and this reduced circuit does not conceal any untestable faults.

The implications of these faults are interesting. In the early days of logic
design, considerable emphasis was laid on logic reduction with a view to minimizing the hardware costs. In recent years, logic minimization has tended to
go out of fashion, on the grounds that silicon is cheap but design effort,
whether measured in terms of human or computer resources, is expensive.

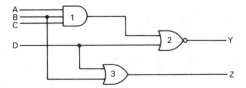

Fig. 2.12 Minimized circuit with the same function as Fig. 2.11.

With an increasing awareness of the cost of testing, however, this approach to logic design may have to be reconsidered.

It should not be assumed that logical redundancy is necessarily the result of poor design: there are at least two considerations that may cause the designer to incorporate redundancy deliberately. The first is the desire to avoid large fan-in gates and to minimize inverters, while restricting implementation to a single type of gate. A well-known example of this is the NAND gate implementation of the exclusive-OR function shown in Fig. 2.13. This implementation is derived by factoring the expression, giving

$$A.\overline{B} + \overline{A}.B = A(\overline{A} + \overline{B}) + B(\overline{A} + \overline{B})$$

Hence, redundant terms have been introduced so as to allow an implementation with four gates rather than the five gates needed for the original expression. In this particular case, however, the redundancy does not result in any untestable faults; although untestability is usually caused by redundancy, redundancy does not necessarily produce untestability.

The second consideration that may lead a logic designer to incorporate redundancy deliberately is the desire to avoid glitches due to unequal propagation delays. The simplest example of this kind is the circuit shown in Fig. 2.14(a) which is an implementation of the function depicted in the Karnaugh map shown in Fig. 2.14(b).

Here the minimal implementation of the function requires only gates G1, G3, G4, and G5; G2 provides the 'bridging' term that prevents an output spike that would otherwise appear when the inputs change from 111 to 110.

If we attempt to generate a test for $D/0$ we find that

a to give $D = 1$ requires $A = B = 1$;
b to transmit the fault through G5 requires $E = F = 0$;
c G3 and G4 must each have at least one input low to give $E = F = 0$; because A and B are both high, the remaining inputs to G3 and G4 must therefore be low.

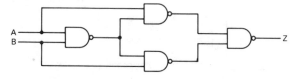

Fig. 2.13 Implementation of the XOR function.

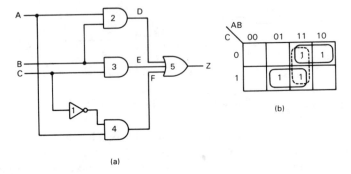

(a)

(b)

Fig. 2.14   Deliberately introduced redundancy. (a) Glitch-free circuit;
(b) Karnaugh map showing bridging term.

d these two signals are logical inverses of each other, hence the requirements are incompatible and the fault is untestable.

Hence we have here a circuit that is not completely testable because of redundancy that has, for good reason, been deliberately introduced. Faced with
this dilemma, there are four possible courses of action open to the designer:

1 Accept that the circuit is not 100% testable, and take a chance that this
   particular fault wil not appear.
2 Remove the redundant element, and with it the untestable fault, and
   take a chance on the glitch.
3 Modify the circuit to make the fault visible. This can be done in various
   ways, the simplest of which is to provide an additional primary output,
   Y, connected to the output of G2, as shown in Fig. 2.15.
4 Modify the circuit to provide additional control over the data
   pathways. Again, there are various ways in which this can be achieved;
   one way, also illustrated in Fig. 2.15, would be to supply an additional
   primary input, *T*, to the input of G4. During normal circuit operation,
   T = 1 and the circuit behaves exactly as it did before. For test purposes,
   we can make *T* = 0, which forces the output of G4 low irrespective of
   the values of *A* and *C*.

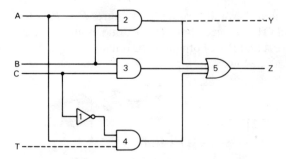

Fig. 2.15   Circuit of Fig. 2.14 modified for increased testability.

It is worth noticing that options (3) and (4) are unlikely to be available to the designer if the existence of the testing problem is noticed only after the circuit has been fully designed. At this stage, extra pathways are likely to be very difficult to accommodate, and extra pins or edge connectors are likely to be even more difficult. Hence, if the full range of options is to be maintained, testing information needs to be available early in the design process so that potential problems can be identified and possible solutions considered before the design has been finalized.

2.5 TESTING FOR BRIDGING FAULTS

Although structurally based test patterns are usually derived largely from the single-stuck-fault model, it is common to expand the fault list by including also some other faults that are believed to be important for that particular circuit. One such fault which is often considered is the bridging fault.

The defect that gives rise to a bridging fault is an unintentional connection between two circuit nodes. On pcbs, bridging faults are most commonly caused by a solder splash joining two adjacent tracks; a similar effect can be produced in an ic, particularly where two layer metallization is used.

A bridging fault differs from a stuck fault in that each of the affected nodes is still able to assume either logic value; if both nodes have a fault-free value of 0, or if both have a fault-free value of 1, then the circuit action is unaffected by the defect. If the fault-free values of the two nodes are different, however, the conflict has to be resolved. The result depends on the technology in which the circuit is implemented; in standard TTL, for example, the low node will dominate, while in ECL the high node will dominate.

As an illustration of the need to include bridging faults in the fault-list, consider the circuit shown in Fig. 2.16. Using the usual techniques, it can readily be established that a complete test set to cover all single stuck faults is

0110/0; 1001/0; 0111/1; 1110/1

It is clear by inspection that this test set will not detect a bridging fault between B and C, since in each of the tests in the set we have $B = C$.

The procedure for generating a test to cover this fault is essentially the same as for stuck faults. We first establish the fault-sensitive condition by requiring $B \neq C$. There are two possible ways of doing this: either $B = 1$ and $C = 0$ or $B = 0$ and $C = 1$. If, for the sake of argument, we suppose that the

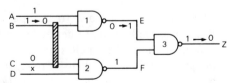

Fig. 2.16 Testing for bridging faults.

circuit is implemented in TTL, then the effect of the fault will be that the line that should be high will be pulled low. The remaining steps in the process follow the usual lines:

a If we choose to establish fault conditions with $B = 1$ and $C = 0$, the fault-effect will be that B will be pulled low.

b To propagate the fault through G1 requires $A = 1$.

c To propagate the fault through G3 requires $F = 1$. This condition is already satisfied because $C = 0$, so that the value of D is immaterial.

There are, therefore, two possible tests derived from the assumption that $B.C = 10$, and a similar argument will produce two more tests for $B.C = 01$. The tests may be expressed as $110X/1$ and $X011/1$.

Consider now a bridging fault between E and F in the circuit of Fig. 2.16. We will require E and F to be unequal, so that $E.F = 10$ or 01, and either way $Z = 1$. Whichever test is chosen, the effect of the fault will be to make $E.F = 00$, and this will also make $Z = 1$. Hence this fault is undetectable.

In considering the choice of bridging faults to be included on the fault list we are faced with another difficulty. We cannot include bridging faults between all pairs of nodes in the circuit; a circuit with a thousand nodes has half a million node-pairs. However, it is clearly unnecessary as well as impracticable to include all these faults, since an accidental short between tracks is hardly likely to occur unless the two tracks are physically adjacent. However, the identification of all possible adjacent tracks in the circuit can be established only by detailed study of the layout; even with a computer-based algorithm this is not easy, and it is likely to consume considerable computer time. A compromise sometimes adopted is to consider just bridging faults between adjacent pins.

SUMMARY
Chapter 2

Among the important features of circuits that need to be distinguished for testing purposes are primary inputs and outputs, internal nodes, fan-out nodes, local and global feedback paths, and positive and negative reconvergence.

Generation of test patterns for logic testing can be carried out on either a functional or a structural basis. The functional approach is difficult to implement, although complex units have to be tested this way.

The structural approach to test pattern generation is based essentially on a pre-defined fault list, derived from a fault model. The single-stuck-fault model, although not sufficiently comprehensive to represent the effects of all defects, is by far the most important, and is widely used as the basis for TPG.

The general strategy of structural testing is to focus on one fault at a time for the purpose of writing a test. Recognizing that any particular test is likely to cover a number of faults apart from the one for which it was specifically written, the fault-cover of each test is checked as soon as it has been generated and the fault-list amended accordingly.

The actual process of test generation first requires the establishment of the fault-sensitive condition and then depends on the principle of path sensitization: the fault has to be transmitted from the site of origin to a primary output at which it can be observed. The essence of this process is that for any particular type of gate the output can be made to depend on one of its inputs by applying suitable values to its other inputs. Almost equally important in practice is fault-blocking, which is the converse of fault-transmission; the ability to make the output of a gate independent of one of its inputs by applying suitable values to its other inputs.

When a circuit contains one or more fan-out nodes, as is very commonly the case in practice, the test generation procedure can become more complicated. In particular, reconvergence demands care both in the generation of tests and in the checking of fault-cover. Positive and negative reconvergence have different effects; but the two types cannot be distinguished on the basis of the circuit topology alone.

It is possible (and, indeed, not at all difficult) to design circuits in which certain faults are undetectable. This is commonly due to the inclusion of logical redundancy in the circuit; and this in turn may be the result of an accidental failure to implement a minimal design, or it may be included deliberately in order to satisfy some other design criterion. There are various possible courses of action open to the designer who discovers the presence of undetectable faults, but the options are severely restricted unless the testing problem is identified at a very early stage in the design process.

Finally, tests for bridging faults can be derived using the sensitive path technique; the only problem that sometimes arises being that some bridging faults are inherently untestable.

EXERCISES
Chapter 2

E2.1 The minimal two-level implementation of the function

$$Z = A.\overline{C} + \overline{B}.C.$$

has a static hazard when the input changes from 101 to 100.
Show that the circuit below

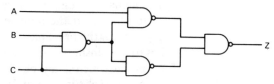

a implements the same function;
b avoids the static hazard;
c can be tested for all single-stuck faults (including faults on the fan-out branches) without requiring access to the internal nodes.

E2.2 Outline the differences between structural and functional approaches to test-pattern generation. Compare the advantages and disadvantages attached to each method.

E2.3 How many tests are needed to cover all single-stuck faults on
a a four-input NAND gate;
b a two-input XOR gate?

E2.4 For the circuit below, show that W/O is undetectable at the primary output. Redesign the circuit so as to produce the same output function without containing any undetectable faults.

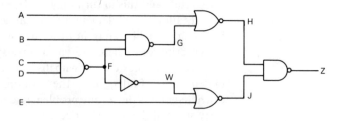

E2.5 The designer of the circuit shown in the diagram has proposed a test set

$$\overline{A}.\overline{B}.C./Z; \; A.\overline{B}.\overline{C}./Z; \; \overline{A}.B.\overline{C}/\overline{Z}$$

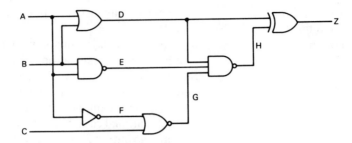

Disregarding possible faults in the fan-out branches:

a determine the fault-cover of the test set;

b generate new tests as necessary to cover the remaining single-stuck faults on nodes A-H and Z.

What faults on the fan-out branches will not be covered by the test set derived above?

E2.6 For the circuit below

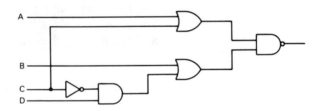

a find all possible tests for $B/1$;

b find all possible tests for $B/0$.

Comment on the relationship between these test sets.

E2.7 The diagram shows a portion of a circuit implemented with TTL gates on a pcb, with P Q R S and Z connected to the edge-connector. When the board goes into production, it is anticipated that the following manufacturing defects may occur:

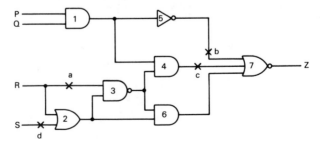

a a short-circuit between point a and the earth rail;

b a solder-splash joining points b and c;

c a break in the track at point d.

Suggest fault-models that will represent these defects; generate tests where possible, and comment on the implications of your results.

E2.8 A combinational circuit has *n* inputs, one of which is *X*, and an output *Z*. The function realized by any such circuit can be written as:

$$Z = X.f + \overline{X}.g$$

where *f* and *g* are functions of all the input variables except *X*.

Show that all possible tests for *X*/1 are given by $\overline{X}\,(f \oplus g)$, and all possible tests for *X*/0 are given by $X(f \oplus g)$.

Could this result provide a reasonable basis for ATPG in a circuit with (say) 30 inputs?

E2.9 State the difference between a defect and a fault in a digital circuit, and explain why there is a need for fault-effect models. Give examples of particular models indicating their strengths and weaknesses.

3

AIDS TO TEST PATTERN GENERATION

3.1 MINIMAL TEST SETS

It will be evident from the examples discussed in Chapter 2 that the TPG process is essentially a long-winded one, involving a considerable amount of data. Whether the TPG is to be performed manually or by a computer, a reduction in the volume of data to be manipulated or in the number of steps in the procedure will be advantageous.

In order to see how TPG can be streamlined, it is helpful first to establish how a minimal test set could be determined, so as to isolate the essential features of the process. We will approach this by making a detailed analysis of a simple combinational circuit.

3.2 THE FAULT-MATRIX

To study the behaviour of a combinational circuit from the point of view of fault-effects, it is useful first to evaluate the fault-coverage of all possible tests on the circuit. The circuit we will consider is shown in Fig. 3.1 which has

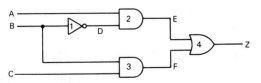

Fig. 3.1 An example circuit.

43

eight possible input test vectors. These will be identified as t_0 to t_7, where the suffix is the decimal equivalent of the binary number (ABC). The circuit has seven nodes, and so fourteen possible faults. The fault-coverage can conveniently be displayed as a table showing by a tick each fault that is covered by each test. The result is shown in Table 3.1; this table is known as a **fault-matrix**.

Table 3.1
Fault matrix for the circuit of Fig. 3.1

	A/0	A/1	B/0	B/1	C/0	C/1	D/0	D/1	E/0	E/1	F/0	F/1	Z/0	Z/1
t_0		✓							✓		✓			✓
t_1		✓		✓					✓		✓			✓
t_2						✓			✓		✓			✓
t_3			✓		✓							✓	✓	
t_4	✓			✓			✓			✓		✓	✓	
t_5	✓					✓	✓					✓	✓	
t_6			✓			✓		✓		✓		✓		✓
t_7			✓		✓							✓	✓	

By examining this matrix we can derive a minimal test-set: the smallest possible set of test patterns that will cover all the faults. This can be done in two stages:

a Any column that contains only a single tick identifies a fault covered by only one test, so that there is no option: if that fault is to be covered, then that test must be included in the test set. Such a test is called an **essential test**; in this example, t_6 is an essential test, because it is the only one that covers $D/1$.

b Having included all essential tests, all faults covered by those tests can be deleted. It is then necessary to find the smallest number of additional tests that will cover all the remaining faults. In this example, t_6 (the only essential test) covers $B/0$, $C/1$, $D/1$, $E/1$, $F/1$, and $Z/1$. The remaining faults can be covered in a number of ways; a suitable set of tests could be written down by inspection, or the full range of options can be derived by writing down a Boolean expression showing the requirements for each remaining fault as indicated in the fault-matrix. This gives

$$(t_4 + t_5)(t_0 + t_1)(t_1 + t_4)(t_3 + t_7)(t_3 + t_4 + t_5 + t_7)$$

which can then be reduced, using normal Boolean algebra techniques, to

$$t_4(t_0 + t_1)(t_3 + t_7) + t_1 t_5(t_3 + t_7)$$

Hence three tests are required in addition to the essential test, and there are six ways in which the tests can be chosen.

It may be observed that this process of obtaining the minimal test set is identical to the Quine–McCluskey process of minimizing a Boolean function (expressed as a sum of minterms) by choosing a set of prime implicants.

It should perhaps be stressed at this point that the use of this method to obtain a minimal test set would never be contemplated in practice: the effort required just to derive the fault-matrix, quite apart from the subsequent manipulations, would be totally uneconomic. For our purposes, however, it serves to expose some features of the process that can be exploited in the interests of simplification.

3.3 FAULT COLLAPSING

The first stage in the TPG process was the formation of the fault-list, which we have taken to consist of each node in the circuit s-a-0 and s-a-1. This body of data has to be stored, referenced, and manipulated during the TPG process; the larger the fault list is, the more computational effort will be required to complete the task. We can, however, reduced the size of the fault-list, and hence the amount of data that needs to be handled, by applying a pre-processing procedure known as **fault collapsing.**

The principle underlying fault collapsing can be demonstrated by observing two typical patterns of fault-coverage that occur in the fault-matrix. These patterns aᵢc illustrated in Table 3.2, which shows a fragment of a fault-matrix relating faults P, Q, R, and S to tests t_i, t_j, t_k and t_1, where these are the only tests that cover these faults.

Table 3.2
Fault matrix patterns that allow fault collapsing

	P	Q	R	S
t_i	✔	✔		
t_j			✔	
t_k	✔	✔	✔	✔
t_l			✔	✔

Consider first faults P and Q. The significant feature of the pattern of ticks in the these two columns is that they are identical; every test that covers P also covers Q, and every test that covers Q also covers P. P and Q are **indistinguishable faults** (represented by $P \leftrightarrow Q$); since any test that covers one also covers the other, there is no need for both faults to appear on the fault list.

Table 3.1 reveals three groups of indistinguishable faults:

$\{A/0, D/0, E/0\}, \{E/1, F/1, Z/1\}, \{C/0, F/0\}$

Our fault-list can be reduced by retaining one fault in each group (say, the first) and discarding the others.

The second pattern in the fault-matrix that permits data reduction is illustrated by faults R and S in Table 3.2. Here we have one fault (S) whose pattern of ticks is a subset of that applying to the other fault (R). In this case we see that if fault S is to be covered we will have to include in the test set either t_k or t_l; whichever one we choose, however, we will at the same time cover fault R. Fault S is said to **dominate** fault R (represented by $S \rightarrow R$), and R can be removed from the fault list. Notice:

a this is an asymmetrical relationship; it must be R that is removed;
b the fault that is removed is the one with the larger number of ticks (although this seems intuitively wrong!).

In Table 3.1 there are several examples of fault-dominance:

$A/0 \rightarrow Z/0; A/1 \rightarrow E/1; C/0 \rightarrow Z/0; C/1 \rightarrow E/1; D/1 \rightarrow B/0, C/1, E/1$

Having removed the dominated faults as well as the indistinguishable faults, the fault-list is reduced to five:

$A/0, A/1, B/1, C/0, D/1$

The reduction in this case is perhaps rather larger than we would normally expect; typically the fault-list can be halved by the use of fault-collapsing.

So far we have looked at fault-collapsing on the basis of data obtained from the fault-matrix. In order for fault collapsing to be useful, however, it is necessary to find a way of applying it without needing to have recourse to the fault-matrix, which, as previously emphasized, will not be available to the TPG system. To see how this can be done, we must think again of the problem of testing isolated gates.

Consider first a two-input AND gate with inputs A and B, and output Z. If we consider the fault $Z/0$, we find that the only test is $A.B/Z$. We will also find that this test covers $A/0$ and $B/0$, and that it is the only test covering these faults. The conclusion is that the group $\{A/0, B/0, Z/0\}$ forms an indistinguisable set of faults.

If now, with the same AND gate, we consider the fault $A/1$, we will find that the only test is $\overline{A}.B/\overline{Z}$, and that this test will also cover $Z/1$. However, it is

Table 3.3
Fault collapsing for basic gates

Type of gate	Indistinguishable faults	Fault dominance
AND	$\{A/0, B/0, Z/0\}$	$A/1, B/1 \rightarrow Z/1$
OR	$\{A/1, B/1, Z/1\}$	$A/0, B/0 \rightarrow Z/0$
NAND	$\{A/0, B/0, Z/1\}$	$A/1, B/1 \rightarrow Z/0$
NOR	$\{A/1, B/1, Z/0\}$	$A/0, B/0 \rightarrow Z/1$
NOT	$\{A/0, Z/1\}; \{A/1, Z/0\}$	None

not the only test to cover this fault: $A.\overline{B}/\overline{Z}$ and $\overline{A}.\overline{B}/\overline{Z}$ also cover $Z/1$. These considerations lead us to the conclusion that $A/1$ dominates $Z/1$.

In a similar way, all the basic gates an be analysed to deduce equivalence and dominance relationships; the results are summarized in Table 3.3. Using these results, fault-collapsing can be performed just on the basis of the circuit description. However, there is one additional complication that needs to be taken into account.

3.4 FAN-OUT NODES

If we apply the relationships in Table 3.3 to the circuit of Fig. 3.1, then we will make the following deductions:

gate 4: $\{E/1, F/1, Z/1\}$ $E/0, F/0 \rightarrow Z/0$
gate 2: $\{A/0, D/0, E/0\}$ $A/1, D/1 \rightarrow E/1$
gate 3: $\{B/0, C/0, F/0\}$ $B/1, C/1 \rightarrow F/1$
gate 1: $\{B/0, D/1\}; \{B/1, D/0\}$ —

If now we check with the full fault-matrix shown in Table 3.1, we find the deductions corresponding to gates 4 and 2 are correct, but those from gates 3 and 1 are not. To investigate the cause of the problem, we will derive tests for $F/0$ and $B/0$, which were predicted above to be indistinguishable.

Following the normal procedure to derive a test for $F/0$ gives:

a $F = 1$ requiring $B = C = 1$
b $E = 0$ to propagate the fault through gate 4.
c Because $B = 1, D = 0$ already and so $E = 0$ irrespective of the value of A; i.e. $A = X$.

Hence the tests for $F/0$ are $X11/1$ (t_3 and t_7).
Similarly, the derivation of a test for $B/0$ gives:

a $B = 1$ to establish the fault-condition.
b The fault can be propagated to the output in one of two ways; either through gate 3 or through gate 2.
  i To propagate through gate 3 requires $C = 1$.
      To propagate further through gate 4 requires $E = 0$. To maintain this value under fault conditions (when $D = 0 \rightarrow 1$) requires $A = 0$.
  ii To propagate through gate 2 requires $A = 1$.
      To propagate on through gate 4 requires $F = 0$.
      Hence $C = 0$.

Thus we have two possible tests for $B/0$: $011/1$; $110/1$ (t_3 and t_6).

These derivations confirm that $B/0$ and $F/0$ are not indistinguishable faults; they also show that there are two features of the circuit that make the faults different:

a t_7, which covers $F/0$, does not cover $B/0$ because of the negative reconvergence which causes the fault-effect at F to be cancelled by another fault-effect at E.

b t_6, which covers $B/0$, does not cover $F/0$ because it uses an alternative sensitive path passing through E instead of F.

Both these effects arise because of the existence of fan-out at node B. Consistency between the true fault implications (as observed in Table 3.1) and those predicted from the individual gates can be restored if we remove all implications relating to the fan-out node; that is to say, those from gate 1 will all be removed, and those from gate 3 will become $\{C/0, F/0\}$; $C/1 \rightarrow F/1$.

This, however is not the end of the story with regard to fan-out. Fig. 3.2 shows a portion of a circuit in which the output of gate 1, labelled A, fans out to input B of gate 2 and input C of gate 3. So far it has been assumed that A, B and C are all part of the same node, and are firmly connected together. Under fault-free conditions, this is, of course, true; but we can easily postulate defects that would make it untrue. If there is a dry joint between the track and the chip pin at B, for example, then the input pin will be open circuit (and hence stuck) while A and C will continue to operate normally, and to be connected together. To allow for this behaviour we can introduce another fault model to distinguish between faults on the **trunk** (A) and faults on the **branches** (B and C). Hence, in Fig. 3.2, we can include A, B and C as separate fault sites, each with two stuck-fault conditions. The understanding here is that faults on A will be transmitted to B and C, but that faults at B will not be transmitted to either A or C.

If fan-out faults are included, then fault-equivalence and fault-dominance relationships associated with the individual gates (as listed in Table 3.3) become valid. As regards fault-dominance relationships between branch and trunk faults, the situation is further complicated by the possibility of negative reconvergence. It was pointed out in section 2.3 that while the existence of reconvergence could be deduced simply from the topology of the circuit, it is not possible to say whether this reconvergence is positive or negative. If reconvergence is present, therefore, no fault-collapsing can be applied between trunk and branch faults. If the fan-out does not reconverge, however, then it is clear that the branch faults (both s-a-0 and s-a-1) will dominate all corresponding trunk faults.

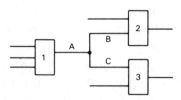

Fig. 3.2 The fan-out problem.

3.5 AUTOMATIC TEST PATTERN GENERATION

3.5.1 Requirements

For a circuit of any size, manual TPG is prohibitively time-consuming; automatic (computer-based) systems are essential. The capabilities of such systems are limited; purely combinational circuits can be dealt with entirely automatically, but sequential circuits present more difficulties to most systems. For reasons that will appear later (see Chapter 8) a system that deals only with combinational circuits is sufficient provided the circuit is designed appropriately.

A number of test generation algorithms have been published, differing in detail but all based on the same idea: tests are generated by establishing sensitive paths in essentially the same way as the manual methods described in Chapter 2. In order to implement this technique, the ATPG system must have access to a variety of data and of routines for the manipulation of data. Among the requirements would be:

a A circuit description, sometimes known as an **image**, specifying the circuit components and their interconnections.
b A fault-list.
c A library of component specifications, derived from the manufacturer's data sheets. These must define
  i   the fault-free behaviour of the component;
  ii the fault-propagation properties of the component.
d A simulation facility for tracing the effects of particular signals (both applied stimuli and fault-effects) through the circuit.
e A means of assessing the fault-cover of a test once it has been generated.

In order to generate a test, the ATPG algorithm will need to perform a sequence of operations which can be summarized as follows:

a **Choose a fault**. This could be simply a random process (taking the faults in the order in which they were listed) or there can be some systematic ordering process. This aspect will be discussed in section 3.5.3.
b **Propagate the fault-effect to a primary output**. This may involve a choice of paths (where there is fan-out) and will certainly involve specifying fixed values on various intermediate nodes.
c **Back-tracking to the primary inputs**. This is necessary to generate:
  i   the fault-sensitive condition at the fault-site (i.e. to test for node N s-a-0 requires the generation of a fault-free value of $N = 1$);
  ii fixed values required to produce the sensitive path.
d **Forward tracking to check consistency**. Once values have been assigned to primary inputs it is necessary to track forward again to ensure that reconvergence effects have been properly accounted for. In particular, it is necessary to consider the possibility of dual (or multiple) path sensitization (see section 2.3).

This alternation of forward and backward tracking through the circuit is an inescapable feature of the TPG process, and it poses a major housekeeping task on the generation algorithm. At all stages of the process, it is necessary to keep making arbitrary choices; for example, if the output of an AND gate is required to be at logic 0, this can be achieved by setting any one of its inputs to logic 0. Having chosen one of the inputs, however, it may turn out that it is not possible to set that particular input to 0, in which case it will be necessary to choose one of the others. Thus the ATPG algorithm must contain:

   a a mechanism for making a choice – this could be arbitrary (say, taking inputs in the order in which they are listed) or could be based on some calculated figure of priority;
   b provision for recording where choices were made so that alternatives can be tried if the first choice fails to generate a test.

3.5.2   D-notation

In order to perform the operations outlined in the previous section, it is necessary to have access to a simulation facility. This facility has to go beyond the normal simulation used in design verification; it has to model not only the fault-free behaviour of the circuit but also the fault-propagation and fault-blocking behaviour of the circuit elements so that path sensitivity can be established or assessed. This calls for an extension of the notation representing logic values beyond the familiar 0, 1, X. The D-notation is widely used for this purpose.

The symbol D is used to represent the state of a node which is at logic 1 under fault-free conditions, but, under the conditions of the particular fault being investigated, is at logic 0. It is, therefore, equivalent to the symbol '1 $\rightarrow$ 0' used in the examples worked in Chapter 2. Similarly, '0 $\rightarrow$ 1' is represented by $\overline{D}$. With the help of this symbolism we can define the behaviour of any particular gate under all the conditions that are relevant to TPG activities. This is illustrated for a two-input NOR gate in Fig. 3.3.

Having settled on a particular fault at a node driven by the output of a gate, we have first to establish the fault-sensitive conditions. The options are specified in Fig. 3.3.(a), which indicates that to establish a fault-free value of 1 at the output of a NOR gate we have to supply 0 to all inputs, while a fault-free value of 0 will be established by applying a 1 to at least one of the inputs. The fault-propagation properties of the gate are defined in Fig. 3.3.(b), which shows that a fault-effect of either kind appearing at one input of a NOR gate will be propagated by making all other inputs 0. A fixed value can be obtained at the output of a NOR gate as shown in Fig. 3.3(c); these same conditions, of course, can also be used to block the transmission of a fault-effect. Finally, the behaviour of the gate under reconvergence conditions is defined in Fig. 3.3(d); the first two illustrations correspond to positive recon-

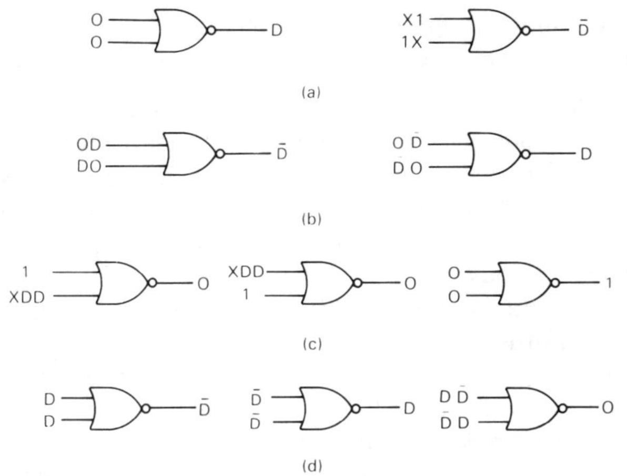

Fig. 3.3 The D-notation as applied to a NOR gate. (a) Establishing fault-sensitive conditions; (b) fault transmission, (c) generating fixed values; (d) reconvergence.

vergence, while the last one corresponds to negative reconvergence.

A similar set of specifications can be drawn up for all the other combinational elements, and these specifications can then be used to formalize the procedures described in section 3.5.1.

3.5.3 Arbitrary choices

Before starting to write a test, it is first necessary to choose which of the faults on the fault-list should be considered. In the examples that have been taken so far, the choice has been made at random, but it is not difficult to see that the order in which the faults are considered can have an effect on the amount of test generation effort that will be expended. The points at issue here can be illustrated by the fault-matrix shown in Table 3.1. If we take the faults in the order shown in the table, we might select tests as follows:

t_5 to cover $A/0$ (also covers $D/0$, $E/0$, $Z/0$)
t_0 to cover $A/1$ (also covers $E/0$, $F/1$, $Z/1$)
t_3 to cover $B/0$ (also covers $C/0$, $F/0$)
t_1 to cover $B/1$
t_2 to cover $C/1$
t_6 to cover $D/1$

Hence we have derived a set of six tests where the minimal test set contains only four tests. The problem is actually caused by the two arbitrary choices that are involved:

a the order in which faults are considered;
b the choice of which test to use to cover any particular fault.

Given the fault-matrix, it is easy enough to see what the best choices would be in each case; without this information the problem is a formidable one, and there is so far no procedure that is guaranteed to identify the optimum choice.

In seeking a minimal test set for 100% fault-cover, it is worth observing that any essential tests will have to be included in the test set. That being so, it would be advantageous to identify the essential tests first, so that all the faults covered by them can be eliminated from the fault-list. This observation can be further generalized by invoking the concept of the difficulty of detecting a fault. A fault is easy to detect if there are many different test patterns that will detect it. Conversely, a fault is difficult to detect if there are very few patterns that detect it. The most difficult fault to detect (apart from undetectable faults, which we cannot do anything about) is a fault for which there is only one test; this test is, by definition, an essential test. If faults can be ordered by difficulty, therefore, there is some reason to suppose that this might be a more efficient procedure than taking random choices.

The application of this principle is hampered by the lack of a rigorous method of assessing the difficulty of a fault without first deriving a fault-matrix. However, faults can be put into a rough ranking order based on empirical observations such as that inputs are more difficult than outputs, and that nodes with fan-out are more difficult that those without. It has been claimed that fault-ordering on this basis makes for economical TPG, but there is no firm evidence that the savings in TPG effort are sufficient to justify the effort necessary to do the ranking.

The other choices within the TPG process are often handled on the basis of a testability analysis of the circuit. This analysis seeks to assess the ease with which individual nodes can be set to defined values, and the ease with which the state of a node can be observed. Testability analysis will be discussed further in section 7.1. Once again, this analysis will be an overhead on the TPG program, and it will therefore have to be justified in terms of the resulting economy in the generation part of the procedure. As a way of keeping this overhead to a minimum, while still providing a basis for choice that is better than a random one, there are some methods of attaching a score to each node, corresponding, for example, to the number of gating levels between the node and the primary inputs or outputs, where the score is easily computed although lacking the subtlety of the more elaborate testability measures.

SUMMARY
Chapter 3

The main problem in TPG for combinational circuits is the volume of data that has to be handled. Some reduction in the size of the fault-list can be achieved by observing that there are characteristic relationships amongst the possible faults in a circuit which ensure that their effects are not all independent. The fault-matrix reveals these patterns, and leads to the definitions of indistinguishable faults and dominant faults. The fault-matrix cannot itself be used for a circuit of realistic size, but a consideration of the properties of individual gates permits most indistinguishable and dominant faults to be identified from the circuit diagram. In this process, particular attention has to be paid to the existence of fan-out nodes and of reconvergence.

The use of fault-collapsing, which significantly reduces the size of the initial fault-list, reduces the volume of the data base, but still leaves the second major problem in TPG. This is the management of the generation algorithm, involving repeated alternations of forward and backward tracking through the circuit in a process that is essentially one of trial and error. The chances of success in this process can be improved by systematic guidance applied to the 'random' choices, but this is at the expense of additional computational effort in testability assessment.

EXERCISES
Chapter 3

E2.2 For a two-input XOR gate find
  a all groups of indistinguishable faults;
  b all examples of fault-dominance.
  Confirm your conclusions by deriving a complete fault-matrix for the gate (4 tests, 6 single-stuck faults).

E3.2 The circuit shown below contains 16 connections (marked A–R), so that 32 single-stuck faults can be defined.

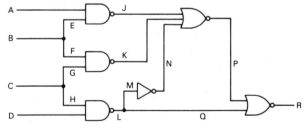

  a How many distinct faults need to be retained on the fault-list for TPG purposes?
  b Develop a test set for the circuit.
  c Is it possible to achieve 100% fault coverage?

Exercises
Chapter 3
continued

E3.3 **a** By applying fault-collapsing to the individual gates in the circuit below, find three sets of indistinguishable faults.
   b By finding all possible tests for $A/1$ and $C/1$, show that these faults are indistinguishable.
   c Show that $B/0$ and $H/0$ are indistinguishable.

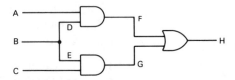

E3.4 The circuit below contains four fan-out nodes, F_1–F_4. For which of these nodes can fault-dominance relationships be correctly deduced?

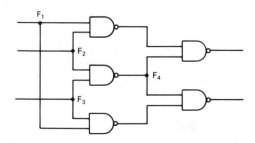

4

FAULT DIAGNOSIS

4.1 FAULT DICTIONARY

The emphasis in Chapters 2 and 3 has been on generating a minimal test set that covers the faults on the fault-list. Such a test set provides a go/no-go decision, which is the first requirement for production testing. However, as pointed out in Chapter 1, there are occasions on which a go/no-go test is not sufficient; if the UUT is to be repaired, diagnostic information is required in addition, so as to locate the fault to the smallest replaceable element in the circuit. Even with an unrepairable unit (an integrated circuit), diagnostic information is useful as an aid to monitoring the performance of the fabrication process.

Access to an ic is necessarily restricted to the pins, and a similar restriction applied to a pcb would simplify the mechanics of applying a diagnostic test program. It is worthwhile, therefore, to consider first the extent to which diagnostic information can be deduced from the results of a test program as observed at the primary outputs. As an example, consider the circuit of Fig. 4.1, in which, for the sake of simplicity, we will not take account of faults on the fan-out branches. The fault-list, therefore, consists of sixteen single-

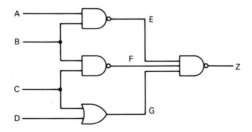

Fig. 4.1 Circuit for which diagnosis is required.

stuck faults. A minimal test set to cover all these faults consists of the five tests

$$\overline{A}.\overline{B}.\overline{C}.\overline{D}/Z;\ \overline{A}.B.\overline{C}.D/\overline{Z};\ \overline{A}.B.C.D./Z;\ A.\overline{B}.C.\overline{D}/\overline{Z};\ A.B.\overline{C}.D/Z$$

which may be conveniently expressed using the notation of chapter 3 (see section 3.2) as

$$t_0,\ t_5,\ t_7,\ t_{10},\ t_{13}$$

In using this test set purely for go/no-go testing, we would follow the procedure indicated in Fig. 4.2(a): if a failure is found on any particular test, the unit is marked as 'no-go' and the test sequence is terminated. This procedure is straightforward, and will maximize throughput on the tester, but provides no diagnostic information whatever. A useful improvement, while keeping the same basic procedure, is to record additionally the point in the test program at which the UUT failed, and, since the circuit will usually have more than one output, the pin at which the incorrect output was observed. This will at least suggest a point at which the diagnostic procedure can start, and is normally used for this purpose in diagnostic systems based on the use of a guided probe (see section 4.2).

Very much more information can be derived from the results of a test program if we adopt the procedure illustrated in Fig. 4.2(b). The essential difference here is that when a failure is observed the test sequence is not truncated. After the sequence has been completed, a record of the result of each

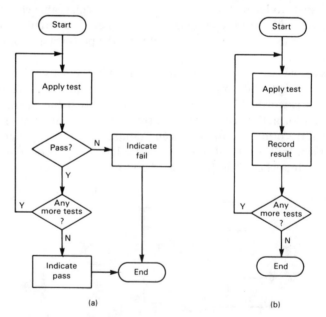

(a)   (b)

Fig. 4.2   Alternative test application procedures. (a) Terminate at first failure; (b) collect complete results.

individual test is available. The diagnostic value of this information can be deduced from a consideration of the fault-cover of each of the tests in the test set. Before starting this, however, it is worth remembering that indistinguishable faults by definition cause identical responses to all possible tests, so that there is no way in which such faults can be distinguished on the basis of test results alone. It will simplify matters, therefore, if we apply fault-collapsing techniques to the circuit before considering the fault-cover of the tests. In the circuit of Fig. 4.1, there are three groups of indistinguishable faults:

$$\{A/0, E/1\}; \{D/1, G/1\}; \{E/0, F/0, G/0, Z/1\}$$

If the diagnostic ambiguity implied by the existence of these groups of indistinguishable faults is unacceptable, the circuit could be modified by the use of test probes on some of the internal nodes, or even by the inclusion of extra input/output pins, in similar ways to those described in Chapter 2 for dealing with undetectable faults (see section 2.4). If the circuit is modified, it then becomes a different circuit, for which is different (probably shorter) test program would be appropriate. The decision to make such changes, therefore, should be made at the earliest possible time if full advantage is to be taken of them.

In the present case, we will assume that no additional access to the circuit is to be provided, so that the circuit can be in one of twelve distinguishable fault states, depending on which (if any) fault is present. The fault-states can be defined as

F_0	Fault-free	F_4	$B/1$	F_8	$\{D/1, G/1\}$
F_1	$\{A/0, E/1\}$	F_5	$C/0$	F_9	$\{E/0. F/0, G/0, Z/1\}$
F_2	$A/1$	F_6	$C/1$	F_{10}	$F/1$
F_3	$B/0$	F_7	$D/0$	F_{11}	$Z/0$

The fault-cover can be displayed in a fault-matrix, as shown in Table 4.1; the information collected during the running of the test, using the procedure of

Table 4.1
Fault matrix for selected tests applied to circuit of Fig. 4.1

		F_1	F_2	F_3	F_4	F_5	F_6	F_7	F_8	F_9	F_{10}	F_{11}
	t_0						✔		✔			✔
	t_5	✔					✔	✔		✔		
Minimal test set	t_7		✔		✔						✔	✔
	t_{10}			✔	✔					✔		
	t_{13}	✔		✔								✔
Additional test	t_1						✔		✔			

Fig. 4.2(b), will be effectively one of the columns of Table 4.1, or it will show that the UUT is fault-free. To the extent that the columns of Table 4.1 are different, therefore, we can, on the basis of the set of test results, perform diagnosis to a fault-group. The diagnostic procedure consists of interrogating the test results as illustrated in Fig. 4.3; with the minimal test set we find that diagnosis to a single fault-group is possible except in the case of faults F_2 and F_7. In order to resolve this final ambiguity it will be necessary to look again at possible tests to cover the faults in question.

The requirement in this case is to find a test for one of the faults that does not cover the other. Investigation will show that $F_2 (A/1)$ is covered only by t_5 which also covers F_7. $F_7 (D/0)$, however, is covered by t_1, t_5 and t_9; neither t_1 nor t_9 covers F_2. The addition of either of these tests to the test set will therefore allow the diagnosis of all distinguishable single-stuck faults just on the basis of test results, without any requirement for access to the internal nodes of the circuit.

The data incorporated in Fig. 4.3, defining how diagnosis can be derived from test results, is known as a **fault dictionary**. The merit of this approach is the fully automatic nature of the procedure, requiring no operator intervention and no special fixturing. There are, however, penalties attached:

a the existence of sets of indistinguishable faults means that diagnosis to a fault-group may not provide diagnosis to the smallest replaceable element;

b the fault dictionary itself represents a very large and complex database; for a circuit with N primary outputs, each test has 2^N possible outcomes, so that the derivation of the fault dictionary is a rather less straightforward process than in the example above.

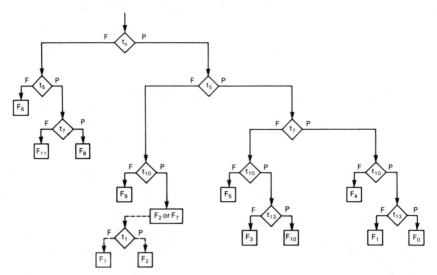

Fig. 4.3 Fault dictionary diagnosis of Fig. 4.1.

4.2 GUIDED PROBE

The main alternative to compiling a fault dictionary for diagnosis is to accept the need for access to the internal nodes of the circuit. It was pointed out in Chapter 1 (see section 1.3.3) that functional testers fitted with edge-connector fixtures are almost invariably provided with a guided probe. This is simply an additional input channel to the tester, which can be connected manually to any node of the circuit. Several features of this facility are worthy of notice:

a The use of a guided probe is only physically possible with pcbs; diagnosis of ics has to rely on a fault-dictionary or similar approach.

b The fact that the probe is connected manually implies that diagnosis is slow and expensive compared with a fault-dictionary approach. On the other hand, it is probably much quicker than fault-finding without the aid of ATE, and certainly requires less skill from the operator.

c With access to internal nodes, most faults that are indistinguishable as viewed only from the primary outputs can be resolved. This is not, however, to say that precise and unambiguous diagnosis is always possible; some of the problems will be mentioned later.

d It is not easy to apply a fault-dictionary approach to other than combinational circuits, but a guided probe can be used equally well with sequential circuits.

The use of the guided probe for diagnostic purposes depends essentially on being able to observe the activity at every node while a test sequence is applied. Diagnosis is performed as an interactive activity, with the operator attaching the probe to various nodes in succession, under the guidance of the test system (hence the name guided probe). In order to be able to provide this guidance, the test system needs access to two sets of data:

a the topology of the circuit;

b the fault-free sequence of activity at each node in response to the test program.

The procedure is quite straightforward: we probe back through the circuit starting from the primary output at which the fault was detected. To evaluate the activity at each node, the circuit must first be initialized, and the test program is then run. The resulting activity at the node being probed is compared with the stored fault-free activity. Having found an element whose output sequence is incorrect, the inputs of that element are then probed. The aim is to identify an element whose inputs are all correct but whose output is wrong.

The procedure can be illustrated by reference to Fig. 4.4, which shows a fragment of a circuit, together with the fault-free sequences associated with each node. Also indicated on Fig. 4.4 is the way in which a particular fault (node N_5 s-a-1) changes activity throughout the circuit, producing an obser-

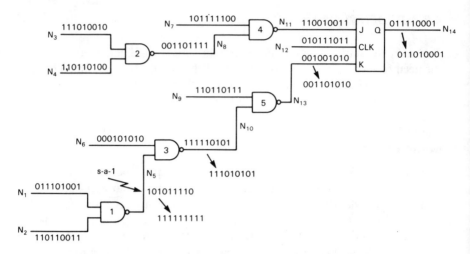

Fig. 4.4 Circuit fragment with associated fault-free and faulty data streams.

vable error at the primary output node N_{14}. Diagnosis in this case might use the following steps:

1 Probe N_{14}.   Incorrect sequence observed.
   Device with output N_{14} has inputs N_{11}, N_{12}, and N_{13}.
2 Probe N_{11}.   Correct sequence observed.
3 Probe N_{12}.   Correct sequence observed.
4 Probe N_{13}.   Incorrect sequence observed.
   Device with output N_{13} has inputs N_9 and N_{10}
5 Probe N_9.   Correct sequence observed.
6 Probe N_{10}.   Incorrect sequence observed.
   Device with output N_{10} has inputs N_5 and N_6.
7 Probe N_5.   Incorrect sequence observed.
   Device with output N_5 has inputs N_1 and N_2.
8 Probe N_1.   Correct sequence observed.
9 Probe N_2.   Correct sequence observed.

At this point we have established that G1 has all correct inputs and an incorrect output; G1 is indicated as most likely to be a faulty element. We should notice that this diagnosis cannot be guaranteed; the fault could be within G1 or within G3 or on the track between. In practice some further refinement is usually possible by probing at both ends of the track between gates; this will expose some defects such as a break in the track, or a dry joint between chip pin and track, but in general some ambiguity will always remain.

The main problem with the scheme as described so far is the enormous amount of data that needs to be stored. A circuit of practical size is likely to have many hundreds of nodes, and the test program associated with it is likely to have many thousands of tests. In assessing the output at each node we are also faced with the problem of performing a comparison (essentially an exclusive – OR function) between two vectors each of several thousand bits length; this is a non-trivial piece of processing. From all points of view, therefore, it is highly desirable that the volume of data that needs to be collected should be reduced.

4.3 DATA COMPRESSION USING TRANSITION COUNTING

The data that is to be compressed, consisting of a long stream of binary data, has the appearance of a signal that changes between two voltage levels at unpredictable times. From the point of view of the external observer the signal is a random one, having no periodicity or any other discernable structure. One simple way in which such a signal might be described is on the basis of the number of times that the signal changes from 0 to 1 or from 1 to 0 during the execution of the test program. This number is the **transition count**; there are several reasons why this measure appears attractive for diagnostic purposes:

a   It is very easy to measure, requiring only a simple counter.
b   A necessary (although not normally sufficient) condition for a circuit to be fully exercised is that under fault-free conditions every node must be required to take each possible value at least once. Hence every node must have a fault-free transition count of at least one.
c   A node that is s-a-0 or s-a-1 will exhibit a transition count of zero, so that such nodes are readily identified.

As well as detecting stuck nodes, the transition count will also be sensitive to many other forms of corruption of the data stream. Transition count was, at one time, widely used in automatic test systems for diagnosis, but it suffers from some serious limitations. Any data compression technique must carry within it a finite probability of mis-classification, since the number of distinguishable compressed measures (often known as **signatures**) is less than the number of different data streams (otherwise we would not have data compression!). There is, therefore, a possibility that the signature of the corrupt data stream can turn out to be the same as the signature of the fault-free data stream. An example of this is to be seen in the circuit of Fig. 4.4, in which, as a result of the s-a-1 fault on node N_5, corrupt data streams appear on nodes N_5, N_{10}, N_{13} and N_{14}. The transition counts for these nodes are shown in Table 4.2: it will be seen that at node N_{13} the sequence changes from 001001010 to 001101010, and for each of these sequences the transition count is the same.

The result of this lack of discrimination is, in this case, disastrous in terms of automatic diagnosis. Following the same procedure as described before, we will observe that N_{14} has an incorrect transition count, and that N_{11}, N_{12} and N_{13} all have correct transition counts. We have therefore found an

Table 4.2
Transition counts for circuit of Fig. 4.4

| Node | Transition count | |
	Fault-free	$N_5/1$
N_5	5	0
N_{10}	4	6
N_{13}	6	6
N_{14}	3	5

element (the flip-flop) with correct input signatures and incorrect output signature, so that we deduce that the flip-flop is faulty. Not only is this diagnosis wrong, but the element identified is not even close the true fault-site.

It has already been pointed out that the possibility of this kind of ambiguity is inherent in the use of any data compression technique; the usefulness of any particular technique will be determined by the probability of a faulty sequence giving rise to the fault-free signature. The faulty sequences that are most likely to give rise to ambiguity are those that differ from the fault-free sequence in only one bit; it will be seen that the problem sequence on node N_{13} in Fig. 4.4 is an example of a single-bit error. The probability of detecting a single-bit error in a long sequence by means of a transition count is only about 50%. The diagnostic performance of transition count is rather better than this when multiple errors are considered, but its poor showing with single-bit errors prompted the development of a more effective data compression technique.

4.4 CYCLIC REDUNDANCY CHECK

4.4.1 Requirements

A data compression scheme for fault diagnosis purposes should yield a short signature whose length is independent of the length of the incoming data stream. The functional requirements can be expressed in terms of the error-detection performance, and include at least two features:

a There should be a high probability that the signature generated by a data stream containing errors will differ from the signature of the fault-free data stream.

b In particular, a data stream containing a single-bit error should have a high probability of detection.

Similar requirements appear in digital communications systems, where the need to provide reliable data transmission is satisfied by incorporating error detection or correction in the data coding. Essentially, such schemes involve expanding each group of data bits by adding check bits in such a way that the expanded group satisfies some predefined mathematical relationship. When a data stream is received, it is tested against this same relationship; failure to satisfy it is taken to indicate the presence of an error.

4.4.2 Modulo-2 arithmetic

Many error coding schemes can be described most usefully in terms of modulo-2 arithmetic, in which addition and subtraction are the same as in normal binary arithmetic with the very important exception that there are no carries or borrows. Addition and subtraction are therefore defined by the basic tables:

$$0 + 0 = 0 \qquad\qquad\qquad 0 - 0 = 0$$
$$1 + 0 = 1 \qquad\qquad\qquad 1 - 0 = 1$$
$$0 + 1 = 1 \qquad\qquad\qquad 0 - 1 = 1$$
$$1 + 1 = 0 \qquad\qquad\qquad 1 - 1 = 0$$

Three observations can be derived from these tables:

a Addition and subtraction are identical:

$$A + B = A - B$$

b In terms of logic implementation

$$A \pm B = A \oplus B$$

i.e. modulo-2 addition or subtraction of two logic signals can be implemented with an exclusive-OR gate.

c The modulo-2 sum of n 1s is equal to 0 if n is even and 1 if n is odd.

Multiplication of two single bits is the same as with ordinary binary multiplication:

$$0 \times 0 = 0 \qquad\qquad 1 \times 1 = 1$$
$$0 \times 1 = 1 \times 0 = 0$$

Multiplying two multi-bit numbers also follows the standard procedure of forming partial products and adding them (modulo-2 of course).

Example 1   Multiply 110111 by 1101

```
        1 1 0 1 1 1    Multiplicand
            1 1 0 1    Multiplier
        _____
        1 1 0 1 1 1 ⎫
      1 1 0 1 1 1   ⎬  Partial
    1 1 0 1 1 1     ⎭  products
    _____
    1 0 1 0 1 0 0 1 1   Product
```

It should be noticed that, since there are no carries, each bit of the product can be formed independently.

It is convenient to represent a data stream as a polynomial in a dummy variable x. Thus the multiplicand in example 1 can be represented as

$$110111 \equiv 1.x^5 + 1.x^4 + 0.x^3 + 1.x^2 + 1.x^1 + 1.x^0$$
$$= x^5 + x^4 + x^2 + x + 1$$

When this is multiplied by

$$1101 = x^3 + x^2 + 1$$

the result is given by

$$(x^5 + x^4 + x^2 + x + 1)(x^3 + x^2 + 1)$$
$$= x^8 + x^6 + x^4 + x + 1$$

where each term in the resultant polynomial is present if that term arises an odd number of times. This corresponds to the straightforward multiplication process:

$$
\begin{array}{llll}
x^5 + x^4 \quad\quad + x^2 + x + 1 & \text{Multiplicand} = D(x) \\
\quad\quad\quad x^3 + x^2 \quad\quad + 1 & \text{Multiplier} \\
\hline
x^5 + x^4 \quad\quad + x^2 + x + 1 & 1.D(x) \\
x^7 + x^6 \quad\quad + x^4 + x^3 + x^2 & x^2.D(x) \\
x^8 + x^7 \quad\quad + x^5 + x^4 + x^3 & x^3.D(x) \\
\hline
x^8 \quad\quad + x^6 \quad\quad + x^4 \quad\quad\quad + x + 1 & \text{Product}
\end{array}
$$

with "Partial products" bracketing the three partial-product terms.

Modulo-2 division also follows essentially the same course as normal division; the result of dividing a dividend by a divisor is a quotient together with a remainder.

Example 2 Divide 101010111 by 1101

```
                        1 1 0 1 1 1      ← Quotient
Divisor → 1 1 0 1 │ 1 0 1 0 1 0 1 1 1    ← Dividend
                    1 1 0 1
                    ───────
                      1 1 1 1
                      1 1 0 1
                      ───────
                        1 0 0 1
                        1 1 0 1
                        ───────
                          1 0 0 1
                          1 1 0 1
                          ───────
                            1 0 0 1
                            1 1 0 1
                            ───────
                              1 0 0      ← Remainder
```

Some important features of modulo-2 division should particularly be noted:

a It follows from the definition of subtraction that any n-digit number can be subtracted from any other n-digit number, leaving a difference with not more than $n-1$ digits. Thus the first line in the division sum shown above represents

   $1010 \div 1101 = 1$, remainder 111

Hence, the number of digits in the remainder is always at most one less than the number of digits in the divisor.

b As in normal arithmetic, the division process can be reversed by multiplication, using the relationship

   Dividend = Divisor × Quotient + Remainder

Thus, taking the data from example 1 used above to illustrate multiplication, we would find that

   $101010011 \div 1101 = 110111$, remainder zero

Comparing this with the division sum above, we see that the two dividends differ in only one bit, and this difference has resulted in a difference in remainder. This sensitivity to changes in a data stream is used as the basis of error coding and of signature analysis.

As with multiplication, division can also be worked out using polynomial representation. Example 2 would then be expressed as:

$$
\begin{array}{r}
x^5 + x^4 \quad\; + x^2 + x + 1 \\[2pt]
\hline
x^3 + x^2 + 1 \;\big)\; x^8 \quad + x^6 \quad\;\; + x^4 \quad\;\; + x^2 + x + 1 \\
x^8 + x^7 \quad\;\; + x^5 \\[2pt]
\hline
x^7 + x^6 + x^5 + x^4 \\
x^7 + x^6 \quad\;\; + x^4 \\[2pt]
\hline
x^5 \qquad\qquad + x^2 \\
x^5 + x^4 \qquad + x^2 \\[2pt]
\hline
x^4 \qquad\qquad\quad + x \\
x^4 + x^3 \qquad\;\; + x \\[2pt]
\hline
x^3 \qquad\qquad\quad + 1 \\
x^3 + x^2 \qquad\;\; + 1 \\[2pt]
\hline
x^2
\end{array}
$$

4.4.3 Hardware realization

The polynomial forms are particularly convenient when considering how modulo-2 multiplication and division can be computed in hardware. The situation with which we are concerned is that in which a string of binary digits is presented serially for analysis. Such a data stream is represented as

$$D(x) = d_0 + d_1 x + \cdots\cdots + d_{n-1} x^{n-1} + d_n x^n$$

$$= \sum_{i=0}^{n} d_i x^i \tag{4.1}$$

If now we suppose that this data stream is presented least significant digit first, the variable x in the polynomial can then be interpreted as a discrete time variable. In other words, data bit d_i becomes available at the ith time instant. If the data stream $D(x)$ is delayed by one time period, each data bit d_i will become available at the $(i + 1)$th time instant; the data stream could then be represented by $x D(x)$. By feeding the data stream to the input of an r-stage shift register, as shown in Fig. 4.5, we will have a number of data streams available:

$$\left.\begin{array}{ll} \text{Input:} & D(x) \quad \ldots + d_{i-1}x^{i-1} + d_i x^i + d_{i+1}x^{i+1} + \ldots \\ \text{Stage 1:} & xD(x) \quad \ldots + d_{i-2}x^{i-1} + d_{i-1}x^i + d_i x^{i+1} + \ldots \\ \text{Stage 2:} & x^2 D(x) \ldots + d_{i-3}x^{i-1} + d_{i-2}x^i + d_{i-1}x^{i+1} + \ldots \\ \text{Stage } r: & x^r D(x) \ldots + d_{i-r-1}x^{i-1} + d_{i-r}x^i + d_{i-r+1}x^{i+1} + \ldots \end{array}\right\} \tag{4.2}$$

These $(r+1)$ data streams can be seen to be the $(r+1)$ possible partial products corresponding to multiplication by an $(r+1)$-bit number. The total product is formed by adding together the appropriate partial products; because of the absence of carries, each bit of the product can be calculated separately. This can be achieved with the circuit of Fig. 4.5 by choosing the states of the $(r+1)$ switches m_0 to m_r. If $m_j = 1$ corresponds to a closed switch, the circuit of Fig. 4.5 computes the product

$$P(x) = D(x) \times M(x)$$

where $\quad M(x) = \sum_{j=0}^{r} m_j x^j \tag{4.3}$

The product will be generated serially, least significant bit first. Inspection of the definitions of the data streams (equation 4.2) shows that at time x^i the output is given by

$$(m_0 d_i + m_1 d_{i-1} + m_2 d_{i-2} + m_3 d_{i-3})x^i = \sum_{j=0}^{r} m_j d_{i-j} x^i$$

Hence $P(x) = \sum_{i=0}^{n} \sum_{j=0}^{r} m_j d_{i-j} x^i \tag{4.4}$

Thus a three-stage shift register with $m_3 = m_2 = m_0 = 1$ and $m_1 = 0$ will compute

$$P(x) = (1 + x^2 + x^3) D(x)$$

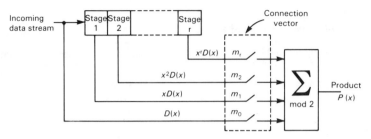

Fig. 4.5   Modulo-2 multiplication using a shift register.

The implementation of division is a somewhat more complicated process. The quotient can be calculated using a shift register with the modulo-2 sum of some of the stages as a feedback function. The input function in this case will be entered most significant bit first, having first initialised the register to the all-zero state. If, for example, we consider the circuit of Fig. 4.6 with the input data stream 101010111 (as in example 2), the results will be as shown in Table 4.3.

Comparison between the 'output' of Table 4.3 and the quotient of example 2 shows that they are the same; the circuit of Fig. 4.6 divides an incoming data stream by 1101. Since the shift register is full of zeros to start with, the quotient will always start with three zeros; with an r-stage shift register the output will start with r zeros. This is correct, since division of a polynomial of order n by a polynomial of order r will be a polynomial of order $(n - r)$.

The complication with a division circuit of this kind is concerned with the remainder. If we repeat the analysis as indicated in Table 4.3 for the input data stream 101010011, we will find that, when the last input bit is entered, the shift register is left with the contents 000. It was shown in example 1 that 101010011 is an exact multiple of 1101, so that the remainder after division by 1101 will also be zero. The last line of Table 4.3, however, shows that when the remainder is not zero (see example 2: the remainder is 100) the quantity left in the shift register, which is called the **residue**, is not necessarily equal to the remainder. It is, in fact, easy to see intuitively that the residue cannot, except in a special case, be equal to the remainder, because the bits inserted into the register (which are not subsequently altered) have to be calculated to be the successive bits of the quotient; they cannot, therefore, be simultaneously the bits of the remainder.

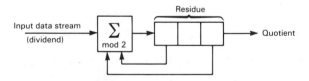

Fig. 4.6   LFSR implementing modulo-2 division of an incoming data stream by 1 1 0 1.

Table 4.3
Operation of division circuit of Fig. 4.6

Input data	Input to shift register	New state of shift register	Output
1	1	100	0
0	1	110	0
1	0	011	0
0	1	101	1
1	1	110	1
0	1	111	0
1	1	111	1
1	1	111	1
1	1	111	1

In any feedback shift register circuit, the value of the divisor polynomial is determined by the pattern of feedback connections. To see how the two are related, it is useful to examine the division process in a little more detail.

In the general case, the divisor will be an $(r + 1)$ digit number given by

$$P(x) = \sum_{i=0}^{r} a^i x^{r-i} \tag{4.5}$$

For simplicity, we will consider division by a four-digit number represented by

$$P(x) = x^3 + a_1 x^2 + a_2 x + a_3$$

The coefficient of x^3 is taken to be 1, otherwise we would be dividing by a number with less than four digits. In the same way we can, without loss of generality, assume that the incoming data stream starts with a 1; it will be represented by

$$D(x) = x^n + d_1 x^{n-1} + d_2 x^{n-2} + \ldots\ldots\ldots$$

For any division to be performed we must have $n \geq r$; normally $n \gg r$. By working through the division process, we can calculate the successive terms in the quotient, denoted by $q_0, q_1, q_2, \ldots\ldots, q_{n-3}$. Notice that the quotient will be of order $n - 3$, so that q_0 is the coefficient of x^{n-3}, i.e.

$$Q(x) = \sum_{j=0}^{n-3} q_j x^{n-3-j} \tag{4.5}$$

The first stage in the division will involve the subtraction

$$
\begin{array}{llll}
x^n + & d_1 x^{n-1} + & d_2 x^{n-2} + & d_3 x^{n-3} + \ldots \\
q_0 x^n + & a_1 q_0 x^{n-1} + & a_2 q_0 x^{n-2} + & a_3 q_0 x^{n-3} \\
\hline
(d_1 \oplus a_1 q_0) x^{n-1} + & (d_2 \oplus a_2 q_0) x^{n-2} + & (d_3 \oplus a_3 q_0) x^{n-3}
\end{array}
$$

Notice that q_0 must be chosen to make the resultant coefficient of x^n equal to zero, in this case, $q_0 = 1$. Notice also that the difference (modulo-2) is represented by the exclusive-OR operation.

From the result of the difference obtained above, it is also clear that

$$q_1 = d_1 \oplus a_1 q_0$$

so that the next stage of the division will be the subtraction

$$
\begin{aligned}
&q_1 x^{n-1} \;+\; (d_2 \oplus a_2 q_0)x^{n-2} \;+\; (d_3 \oplus a_3 q_0)\, x^{n-3} \;+\; d_4 x^{n-4}\\
&q_1 x^{n-1} \;+\; a_1 q_1 x^{n-2} \;+\; a_2 q_1 x^{n-3} \;+\; a_3 q_1 x^{n-4}\\
\end{aligned}
$$

$$\overline{\qquad\qquad [(d_2 \oplus a_2 q_0) \oplus a_1 q_1]x^{n-2} \;+\; \ldots\ldots}$$

Hence

$$q_2 = [(d_2 \oplus a_2 q_0) \oplus a_1 q_1]$$

and by repeating the process again

$$q_3 = \{[cd_3 \oplus a_3 q_0] \oplus a_2 q_1] \oplus a_1 q_2\}$$

The general term in the quotient can now be seen to be

$$q_r = d_r \oplus a_1 q_{r-1} \oplus a_2 q_{r-2} \oplus a_3 q_{r-3}$$

and this expression can be implemented by the circuit of Fig. 4.7, where the switches indicate that the feedback connection is made if $a_r = 1$ and omitted if $a_r = 0$. Hence, we can confirm that the circuit of Fig. 4.6 will divide an incoming data stream by 1101.

4.4.4 Error detection performance

In using a division circuit as a data compression device we rely on the residue as a representation of the data stream. This residue is called the **cyclic redundancy check** (CRC) signature, and it will be useful to the extent that different data streams generate different signatures. When discussing transition counting it was pointed out that its performance was poor particularly when faced with a data stream containing only a single error; the probability of detecting such an error by transition counting is only about 50% for a long data stream. With the CRC signature, however, it will be seen that a single error is certain to be detected. This can be shown by considering the effect of a single error in the circuit in Fig. 4.7:

a Because of the properties of the modulo-2 addition circuit, an error in a particular data bit (d_r say) wil produce an error in q_r.

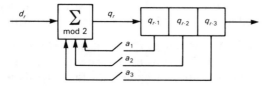

Fig. 4.7   General form of LFSR to implement modulo-2 division.

b The value of q_r is not subsequently changed, but is simply shifted through and eventually disappears off the end. In itself this does not help in diagnosis, as the quotient is not observed when using signature analysis.

c As the incorrect q_r value is shifted through the register, it reaches one of the stages that is fed back to the input, and, in the absence of any other error in the data stream, this must generate a further error in the q stream.

d By the time the original error reaches the end of the register, the q stream will contain several errors due to the feedback. Some cancellation of errors may therefore occur, but it can never be possible to lose all the errors in this way, since cancellation requires at least two errors in the register. When the first of these is shifted out we are then left with only one, which is therefore bound to be regenerated.

It will clearly be possible for the correct signature to be generated by an incorrect data stream, but only if the data stream contains multiple errors, and only if the errors are arranged in particular patterns. This can be illustrated by reference to Fig. 4.6. If d_r is wrong, q_r will be wrong. This is fed back to form q_{r+1}, but its effect would be cancelled if there were an error in d_{r+1}. It is fed back again at time $(r+3)$, and this would be cancelled if there were an error in d_{r+3}. Hence identical signatures will be generated by the two data streams

$$\ldots . d_{r-1}, \quad d_r, \quad d_{r+1}, \quad d_{r+2}, \quad d_{r+3}, \quad d_{r+4}, \ldots .$$

$$\ldots . d_{r-1}, \quad \overline{d_r}, \quad \overline{d_{r+1}}, \quad d_{r+2}, \quad \overline{d_{r+3}}, \quad d_{r+4}, \ldots .$$

The particular error-patterns that are undetectable are determined by the feedback connections and hence by the value chosen for the divisor. In practical implementations of signature analysis, whether as a stand-alone instrument or as part of a testing system, the choice of feedback connections is governed by two main considerations:

a The free-running feedback shift register (that is, with no input data applied) can be made to cycle through all possible states. This feature will be discussed in Chapter 9 (see sections 9.2.2 and 9.3.2); it is not actually necessary for the efficient functioning of the signature analyser, but it is convenient in a multi-purpose application such as a self-testing system.

b Regularly spaced feedback taps are avoided, particularly four- or eight-bit spacing, because of the likelihood that a fault in a bus-structured system could produce multiple data stream errors with that spacing.

An estimation of the performance of this system can be made by considering an r-bit shift register supplied with an n-bit data stream. The total number of possible data streams is 2^n, of which one is the fault-free data stream, so that the number of incorrect data streams is

$$N_i = 2^n - 1$$

The total number of possible signatures is 2^r. Assuming that the conversion of data streams to signatures is uniform, we can say that each signature corresponds to 2^{n-r} data streams, of which one is the fault-free one, so that the number of incorrect data streams that produce the fault-free signature is

$$N_u = 2^{n-r} - 1$$

There are, therefore, N_u incorrect data streams that the signature analyser fails to detect, out of a total of N_i incorrect data streams that are possible. The probability of an undetected error in a long data stream ($n \gg r$) is therefore given by

$$P_e = \lim_{n \to \infty} \frac{N_u}{N_i} = \lim_{n \to \infty} \frac{2^{n-r} - 1}{2^n - 1} = 2^{-r}$$

For an eight-stage shift register, therefore, the probability that an incorrect data stream will escape detection is about 0.4%, while for a sixteen-stage register it is about 0.002%. These probabilities, based on the assumption of an infinitely long data stream, are actually worst-case estimates; for finite-length data streams performance improves with decrease in length until, for $n \leqslant r$, an incorrect data stream is certain to be detected. (In this case, of course, there is no data compression!) This level of performance of the CRC signature has led to its widespread adoption as a diagnostic tool.

4.4.5 Tri-state components

Any method of data compression, whether by transition counting or by CRC signature, cannot be expected to produce useful results if the fault-free data stream at any node is inconsistent. In this respect, tri-state bus architectures, which are commonly encountered in modern digital circuits, pose a problem to the test programmer in that, if a bus line enters its floating state at any time during the test program, the logic value at that time will be indeterminate. As far as the test program itself is concerned, this is easily dealt with since there is invariably provision in the tester for any particular output to be disregarded (masked) at any particular time. In a signature collection system, however, this would be much more difficult to arrange; it would require different nodes to be treated differently during data capture, which would destroy the essential simplicity of the method.

There are two possible ways round the problem:

a By connecting pull-up (or pull-down) resistors to every bus line we can ensure that the line assumes a definite value when it is not otherwise driven. This, however, uses extra components, and also slows down bus transitions, which could in itself lead to inconsistent signatures.

b The data collection probe (guided probe or inbuilt sensor) can be pulled to a mid-threshold value (about 1.4 V for TTL) through a high resistor, and the signal sensed through a Schmitt circuit that provides substantial hysteresis. In this way we ensure that the value sensed can change only if the node is actively pulled past the hysteresis; a floating state leaves the sensor reading the last firm value.

4.5 THE LOOP-BREAKING PROBLEM

The principle of probing through a faulty circuit until an element is found with correct inputs and wrong output seems straightforward enough, and in most situations will provide a good indication of the fault location. There is, however, one situation in particular in which this approach fails totally.

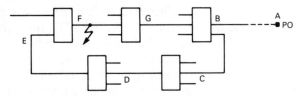

Fig. 4.8 The loop-breaking problem.

Figure 4.8 shows a fragment of a circuit incorporating a global feedback path. If this circuit contains a defect that causes the data stream at node F to be faulty, the fault-effect will be propagated through G and B and eventually to the primary output at A where it will be detected. It will also be propagated through the feedback path to nodes C, D and E. In the diagnosis phase we will work back from the output until we get into the loop; we then find that each of the components in the loop has both wrong outputs and wrong inputs, so that diagnosis cannot do better than list all the components in the loop as possible fault-sites. This is the loop-breaking problem; some research effort has been devoted to the solution of this problem, but no effective procedure has so far been published. The only practical answer is to make provision for global feedback loops to be broken for testing purposes. This point is discussed further in Chapter 7 (see section 7.4.1).

SUMMARY
Chapter 4

Fault diagnosis is required so that a defective element in a circuit can be identified with a view to repair, or to provide feedback data for improving the yield of the manufacturing process. Diagnosis can be performed on the basis of a fault dictionary. It is usually possible in principle to achieve complete diagnosis for a combinational circuit to the smallest element of interest, but there are cost penalties to this approach:

a A large diagnostic data base has to be stored with the program, and provision has also to be made for storing a large volume of test result data collected during the running of the test program.

b In addition to the minimal test set, further tests will normally be needed if all diagnostic ambiguities are to be resolved.

c Diagnosis to the smallest element of interest may require the observability of the circuit to be enhanced by incorporation of additional test points.

The main method of diagnosis as applied to pcbs is by use of a guided probe. The principle here is to seek a component whose outputs are wrong although its inputs are correct. This can be assessed by having a stored representation of the fault-free data stream generated at each node in the circuit in response to the test program. The quantity of data this represents means that the method is not feasible without the use of some kind of data compression. Transition counting has the merit of simplicity, but is not sufficiently sensitive to variations in the data stream.

The cyclic redundancy check is based on the use of a feedback shift register, whose operation is equivalent to modulo-2 division by a particular polynomial defined by the pattern of feedback connections. A single error in the data stream is certain to be detected by such a system, and the probability of failing to detect a multiple-error data stream can be made as small as we like by increasing the length of the shift register.

The presence of global feedback can defeat the diagnostic process by causing erroneous data streams to appear at all nodes round the loop. This problem can be avoided only by making design changes.

EXERCISES
Chapter 4

E4.1 For the circuit shown, establish a fault matrix relating the 16 possible tests to the 20 single-stuck faults. Hence deduce a minimal test set for the circuit. Compile a fault dictionary for this test set and so establish to what extent diagnosis to gate

Exercises
Chapter 4
continued

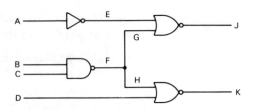

level can be performed without internal probing. (Note: each test has four outcomes corresponding to the correct/incorrect values on the two outputs.)

E4.2 What is the purpose of signature analysis? Explain how it is used to assist in testing.

Describe two forms of signature that have been used, and discuss the advantages and disadvantages of each.

E4.3 What are the advantages and disadvantages of a fault dictionary approach to fault-location, as compared with the use of signature analysis?

E4.4 Evaluate the modulo-3 division $X \div Y$, where

$$X = 1\,0\,1\,1\,0\,1\,1\,0\,1$$
$$Y = 1\,1\,0\,1$$

If the result of this calculation is a quotient, Q, and a remainder, R, show that

$$X = Q \times Y + R$$

(where all arithmetic operations are modulo-2).

E4.5 Design a circuit which will perform modulo-2 multiplication of a serially-presented data stream by $1\,0\,1\,0\,1$.

E4.6 The signature analyser shown below is initialized to the all-zero state, and is fed with the data stream $1\,0\,1\,1\,0\,1\,1\,0\,1$ (presented MSB first).

a What is the fault-free signature?

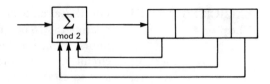

    b Show that if the third bit alone is in error the final signature is different.

    c Find a sequence starting with two correct bits and an error in the third bit, which will generate the fault-free signature.

    d Derive the state-transition diagrams and verify the results of **b** and **c**.

E4.7 With a 6-bit signature register, what is the probability of failing to detect an erroneous data stream of length

    a 100 bits;

    b 10 bits;

    c 5 bits?

5

TESTING SEQUENTIAL LOGIC

5.1 CHARACTERISTICS OF SEQUENTIAL CIRCUITS

The essential feature of a sequential circuit that distinguishes it from a combinational one is that the result of applying any particular set of inputs depends not only on those inputs but also on the current state of the circuit. Hence, the operation of such a circuit cannot be specified from a simple truth table. From the testing point of view this makes for a number of complications.

   a To conduct a single test may require the application of a sequence of patterns in order to establish the appropriate state before applying the required inputs.

   b The truth table of a combinational circuit constitutes a complete functional description of the circuit, so that an exhaustive test set is, in principle, easily specified. To produce a similarly complete functional description of a sequential circuit requires a specification of the outputs in response to all possible sequences of inputs.

   c Because the sequence of input changes is significant, the tester will need to take account of timing information as well as the required logic levels. Even small timing differences, such as are introduced by transmission along signal lines of different lengths, can bring about changes in the circuit's response.

In the light of these difficulties, it is not too surprising that the generation of test patterns for sequential circuits tends to be more of an art than a science. The ideas presented here cannot therefore be regarded as definitive, but only as one possible approach to the problem.

5.2 FUNCTIONAL DESCRIPTION OF A FLIP-FLOP

5.2.1 Mathematical model

Some of the difficulties in forming an adequate functional description of a sequential circuit are well exemplified by a consideration of just a single flip-flop. Textbooks on logic design usually define the behaviour of flip-flops by equations such as

$$Q^+ = J.\bar{Q} + \bar{K}.Q$$

and

$$Q^+ = D$$

(5.1)

where Q is the 'present state' and Q^+ is the 'next state'. If, however, we look at practical implementations of flip-flops, we find that there are at least three quite different forms of the circuit, and that the equations above do not adequately describe any of them; in all cases they are imprecise, incomplete and inaccurate.

The imprecision in the definition is bound up with the meanings of 'present state' and 'next state'; this attempt to describe the sequential nature of the device is usually said to refer to two time instants separated by a clock pulse. This, however, conceals an important facet of the behaviour of the flip-flop; the output depends not only on the data inputs but also on the clock input, and especially on the time relationships between changes on clock and data inputs. The nature of this time-dependence is different for the different forms of implementation of the flip-flop.

The simplest form of clocked flip-flop, shown in Fig. 5.1, is the D-type latch; this circuit is the central element in level-sensitive scan-design methodology, which will be discussed in Chapter 8, and is available as a discrete component (e.g. 7475). In a level-sensitive flip-flop circuit, the output responds immediately (apart from gate propagation delays) to changes of data input whenever the clock is high; and while the clock is low, the output remains stable. This behaviour is shown diagrammatically in Fig. 5.2, in which the time-varying data inputs are denoted by $I(t)$, and $f[I(t)]$ is the functional relationship between output and input.

Edge-triggered and master/slave flip-flops permit output changes only on clock pulse edges. D flip-flops are normally edge-triggered, but JK flip-flops are available in both forms; the timing characteristics are illustrated in Fig. 5.2 for a positive-edge-triggered flip-flop (e.g. 74109 ($J\bar{K}$) or 7474 (D)) and

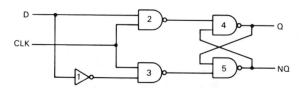

Fig. 5.1   Circuit diagram of a clocked D-type latch (level-sensitive flip-flop).

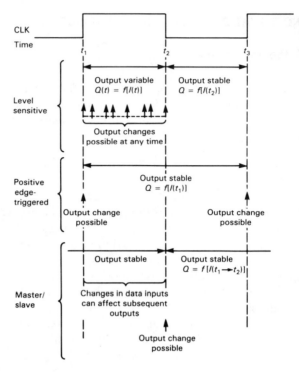

Fig. 5.2   Time-responses of the three main types of flip-flop.

for a master/slave (e.g. 7476 (*JK*)). Notice particularly that a *JK* master/slave flip-flop has sensitivity to input changes while the clock is high, even though the resulting output changes do not appear until the negative transition of the clock. Specifically, if, while the clock is high, a glitch appears on either the *J* or *K* line when both are nominally low, the master will be set or reset; this is the '1s catching' property of the flip-flop. This sensitivity is represented in Fig. 5.2 by the notation $Q = f[I(t_1 \rightarrow t_2)]$, indicating that inputs between t_1 and t_2 need to be taken into account when computing the output.

When flip-flops and latches are incorporated into a logic circuit, the differences in detail outlined above may or may not be significant in terms of circuit performance, depending on the design methodology used. The test engineer needs to be aware of the possible complications that are not represented by eqns (5.1). This requires detailed knowledge of the particular components used; a task that is not made any easier by some manufacturers who do not specify on the data sheet whether a given device is edge-triggered or master/slave! The complications become particularly evident if the signal applied to the clock input is a circuit variable (rather than a master clock signal obtained directly from a primary input), or if the flip-flop appears within a global feedback path; they can be avoided by sticking to a strictly synchronous design methodology. This aspect will be discussed further in Chapters 7 and 8.

5.2.2 Asynchronous behaviour

A further complication with commercially available flip-flops is the provision of one or more asynchronous inputs. Figure 5.3 shows the symbol used by one particular manufacturer for a 74112, which is a negative-edge triggered JK flip-flop; it may be noted in passing that exactly the same symbol is used by the same manufacturer for the 7476, which is a master/slave flip-flop. This emphasizes again the need to go beyond the circuit diagram when assessing the testing needs of the circuit. The inputs S_D and C_D would be applied to gates 4 and 5 in the circuit of Fig. 5.1; the important feature of these inputs is that they by-pass the clock, producing output changes asynchronously. This, of course, further obscures the meaning to be attached to Q^+, and also introduces the possibility of making both asynchronous inputs active simultaneously. Under these conditions, instead of having the normal state where $NQ = \overline{Q}$, we have instead $NQ = Q = 1$.

In an attempt to represent the complete behaviour of this flip-flop, we could consider by the pair of equations

$$\left.\begin{array}{rcl} Q^+ & = & [(J.\overline{Q} + \overline{K}.Q).\ CP + Q.\overline{CP}]S_D.C_D + \overline{S_D} \\ NQ^+ & = & [(\overline{J}.\overline{Q} + K.Q).\ CP + Q.\overline{CP}]S_D.C_D + \overline{C_D} \end{array}\right\} \qquad (5.2)$$

where CP is taken to represent 'a negative edge has been applied to the CP input'. This description avoids the awkward question of what is meant by Q^+, but, although in some ways closer to the truth than eqn (5.1), is still not complete; in particular, it does not account for the indeterminacy of operation when S_D and C_D make a simultaneous transition from low to high. As with the differences in clocked behaviour between different types of flip-flop, the anomalies in response to the asynchronous inputs become particularly evident when internally generated signals are used to drive these inputs. Once again the test engineer needs to be wary in these circumstances.

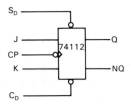

Fig. 5.3   Symbol for the 74112 : a negative-edge-triggered JK flip-flop.

5.3 STRUCTURAL APPROACH TO TESTING A FLIP-FLOP

5.3.1 Forming a fault-list

To illustrate the problems involved, we will develop a test procedure for the 74112 shown in Fig.5.3. If we are to adopt a structural strategy, we must first define a set of faults; in the absence of anything better, these can be based on the single-stuck fault-model. It would, in principle, be possible to use the internal structure of the flip-flop as the basis for generating a test set by the usual process of establishing sensitive paths through the circuit to the outputs. The validity of such a procedure is somewhat questionable, because the internal structure of any integrated circuit may or may not conform to the schematic shown in the data sheet; the manufacturer usually describes this schematic as a functional equivalent. Hence, a program generated on this basis may well contain tests for faults on nodes that do not exist.

One set of nodes that certainly does exist is the set of inputs and outputs; faults on these nodes are known as **pin-faults**. The fault-list to be used as the basis for this programme will be the set of pin-faults:

$$S_D/0, S_D/1, J/0, J/1, CP/0, CP/1, K/0, K/1, C_D/0, C_D/1,$$
$$Q/0, Q/1, NQ/0, NQ/1;$$

that is, each pin stuck at each value.

5.3.2 Initialization

It is important to realize that when a sequential circuit is first switched on it is, in general, impossible to predict what state the circuit will settle in. The first requirement, therefore, is to put the circuit into a known state before the test program can be started.

In the present example there is no difficulty, since the circuit is provided with, and we have access to, the asynchronous inputs S_D and C_D, which give unconditional control over the outputs. An isolated flip-flop will, in any case, normally present no initialization problem (apart from remembering to do it!) but the same is not necessarily true, as will become apparent later, when the flip-flop is embedded in a circuit.

5.3.3 Asynchronous tests

Having initialized the flip-flop, which is obviously best done with the use of either S_D or C_D, it is then convenient to complete the tests necessary to detect all the faults associated with the asynchronous inputs. Notice that as far as these inputs are concerned the circuit is essentially combinational; the output

Table 5.1
Asynchronous tests for a *JK* flip-flop

Test	Inputs	Fault-free output	Faults covered
1	$S_D = 1, C_D = 0$	$Q = 0, NQ = 1$	$Q/1, NQ/0, S_D/0$
2	$S_D = 0, C_D = 1$	$Q = 1, NQ = 0$	$Q/0, NQ/1, S_D/1, C_D/0$
3	$S_D = 1, C_D = 0$	$Q = 0, NQ = 1$	$C_D/1$

in each case is dependent only on the current inputs and not on the previous state.

For the asynchronous tests we will hold *J*, *K* and *CP* all low; this is an arbitrary choice, since the values of these inputs will not affect the outputs. Three tests are then sufficient to cover all pin-faults on the asynchronous inputs, and these will also cover output pin faults. This segment of the program is summarized in Table 5.1. The only comment necessary is that we are assuming here that test 1 is the initialization pattern; for this reason, $C_D/1$ is not covered by that test because the flip-flop might settle in the fault-free state on switch-on without being forced by C_D.

5.3.4 Clocked tests

Having completed the asynchronous tests, we will hold S_D and C_D at their inactive level (high) and then exercise the rest of the circuit by use of the clock.

The test conditions necessary to expose faults on *J* and *K* can be deduced from Table 5.2, which shows the fault-effects resulting from each single-stuck fault. The blank entries in the table indicate those conditions in which the fault-free and faulty inputs are the same, and which cannot therefore constitute a test for the corresponding fault. In forming this table, it has been assumed that a complete clock pulse is applied after the appropriate input conditions have been established. If this is so, Q^+, the output after the clock pulse, obeys the basic state table (as represented by eqn (5.1)) and will be the same irrespective of the type of flip-flop (level-sensitive, edge-triggered, or master/slave).

To derive a test for any particular fault we must choose the state and the inputs such that the fault-free and faulty outputs are different. The available

Table 5.2
Effects of input faults on a *JK* flip-flop

	Input conditions			Q^+			
	J	*K*	Fault-free	*J*/0	*J*/1	*K*/0	*K*/1
1	0	0	Q		1		0
2	0	1	0		$\bar{Q}$	Q	
3	1	0	1	Q			$\bar{Q}$
4	1	1	$\bar{Q}$	0		1	

options can be deduced from Table 5.2. To detect $J/0$, for example, we need to use either line 3 or line 4. In line 3 we would need to ensure that the fault-free output (1) differs from the faulty output (Q): this requires $Q = 0$. Similarly, if we use line 4, we will need to choose Q so that $\overline{Q} \neq 0$, i.e. $Q = 0$.

Continuing in this way we will deduce that:

a to test for either fault on J requires $Q = 0$, and the value of K is immaterial;

b to test for either fault on K requires $Q = 1$, and the value of J is immaterial.

Hence a minimum of four tests is needed; we anticipate that the number of input changes necessary may well exceed this minimum because of the requirement for the flip-flop to be in a specific state for each test. The number of these extra 'setting-up' test vectors can be kept to a minimum by a careful choice of the order of consideration of faults.

If we assume that the flip-flop has first been subjected to the test sequence shown in Table 5.1, its state at the start of the clocked test sequence will be $Q = 0$. We should therefore start with a test for one of the faults on J, say $J/0$. This test is shown in line 1 of Table 5.3; it results in a change of state to $Q = 1$. We may notice in passing that this test will also cover both possible faults on CP, since the output will not change without the application of a clock edge. Because now $Q = 1$, the next fault to cover will be one of the faults on K, say $K/0$. This is shown in line 2 of Table 5.3; it again results in a change of state to give $Q = 0$. In line 3 the other fault on J is covered, leaving the state at $Q = 0$. At this point, we are left with only one fault ($K/1$) still to be covered, but the flip-flop is in the wrong state to test it. Line 4 therefore is inserted simply to change the state, and line 5 is the test to cover the final fault.

The total test sequence, covering both asynchronous and clocked tests, is illustrated in the timing diagram in Fig. 5.4. Two points are worthy of notice:

a In the asynchronous sequence, the transition between lines 1 and 2 of Table 5.1 is handled in two stages: C_D is brought high before S_D is brought low. This avoids the race inherent in making nominally simultaneous changes at these terminals.

Table 5.3
Synchronous tests for a JK flip-flop

	Test for	Initial state Q	Inputs J	K	Fault-free output Q^+
1	$J/0$	0	1	0	1
2	$K/0$	1	1	1	0
3	$J/1$	0	0	1	0
4		0	1	1	1
5	$K/1$	1	1	0	1

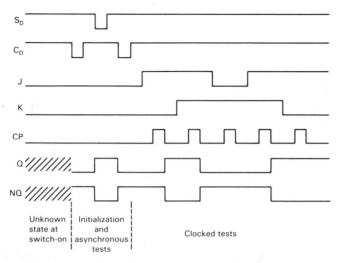

Fig. 5.4   Timing diagram showing a test sequence for a 74112.

   b In each test the logic inputs are established first, and the clock activated afterwards. This avoids all timing problems, including dependence on set-up and hold times.

One final point can be deduced from a study of Table 5.3. The choice of which fault to cover on lines 1 and 2 appeared to be arbitrary. However, further consideration of the response of the flip-flop to each test will reveal that by changing the order in which the faults are covered, it is in fact possible to complete the task using only four tests.

5.4   TESTING SEQUENTIAL CIRCUITS

5.4.1   Choosing the strategy

A sequential circuit, consisting of a number of flip-flops embedded in a mesh of combinational logic, presents problems very similar to those already encountered in the more complex of the combinational circuits. These problems are concerned essentially with the ease of access of the tester to the circuit; the flip-flop states will have to be set up and observed indirectly when combinational logic (or another flip-flop) is interposed between the unit being considered and the input/output pins. This aspect of the problem will be considered further in Chapter 7 (see section 7.1.2).

   To demonstrate a possible approach to the test-pattern generation problems of sequential circuits we will consider the circuit of Fig. 5.5. This is a circuit with three T-type flip-flops: in practice a T flip-flop would probably be a JK flip-flop with J and K shorted together to form the T input.

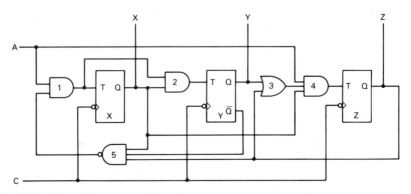

Fig. 5.5 A simple sequential circuit.

In considering the strategy to be adopted in testing this circuit, we are faced immediately with a difficulty. Functional testing, centred around the idea of checking that the circuit as a whole does what it is supposed to do, would be an attractive proposition, but it is made difficult or impossible if the function of the circuit is unknown to the test engineer. Ideally, the circuit designer should himself write the test program, in which case functional testing would clearly be appropriate, but if the two operations are separate then the test engineer will not normally have any data apart from the circuit diagram, from which the function is certainly not obvious. With a circuit the size of figure 5.5, it would, of course, be possible to work out a state transition diagram and then to use that as a functional specification. With a circuit of a realistic size, however, this is not feasible. Even a modest increase in complexity will cause a dramatic increase in the effort required: a circuit with 6 inputs and containing 10 flip-flops – still a very small circuit occupying probably no more than a dozen chips – will require over 65000 state transitions to be evaluated. Even if this were done, it is very doubtful whether the resulting complex mass of data would be any easier to interpret in functional terms than the original circuit diagram.

In view of the difficulties attaching to a functional strategy, and of our understandable reluctance to accept a full structural strategy, we are driven to adopt a hybrid approach in which we first attempt to define and test the function of major elements of the circuit and then add further tests to complete coverage of the interconnecting gates on a structural basis. What constitutes a 'major element' is really a matter for the experience and judgement of the test engineer: any group of circuit elements that together form a recognisable functional unit can be treated as a single entity and subjected to a functional test. This would simplify go/no-go testing, although it is worth pointing out that diagnosis with such a program could not do more than identify the unit as faulty. A complex chip, however, can very reasonably be treated as a single entity; the manufacturer's data sheet can be used to provide the functional description.

The circuit of Fig. 5.5 will, therefore, be tackled on the basis of first applying functional tests to the flip-flops. In the course of this exercise, it is likely that much of the testing needed for the interconnecting gates will be covered fortuitously; the fault-cover will be evaluated after each test, and any gaps that are left after the functional testing is complete will be filled afterwards.

5.4.2 Functional test for the flip-flop

As soon as we start trying to devise a suitable functional test sequence for the flip-flops, we are faced with another practical problem. Figure 5.5 shows, for convenience, T flip-flops as shown in Fig. 5.6(a). However, if such elements are not available the circuit may actually consist of JK flip-flops type 74112, connected as shown in Fig. 5.6(b). The question then is: how should this device be tested? One possibility to be considered could be the test sequence developed in section 5.3, but this is not possible in the circuit as it stands, because

a we do not have access to NQ, so that tests that rely on errors observed at that output will be nullified;
b we do not have access to S_D and C_D, so that tests that rely on manipulation of these inputs cannot be performed;
c we do not separate access to J and K, so that we cannot test the correct functioning of the flip-flop for the conditions in which $J \neq K$.

At first sight, these problems seem to demand modifications to the circuit so as to allow the additional access needed to permit a full test of the device. Further consideration, however, suggests an alternative interpretation. If the circuit being tested does not use some particular performance feature of a device, it can be argued that there is really no point in testing whether that feature is working. An unused feature should, therefore, be ignored for testing purposes; in this particular circuit, Fig. 5.6(a) is, in fact, a better representation of the 74112 than Fig. 5.6(b). It is not uncommon for standard

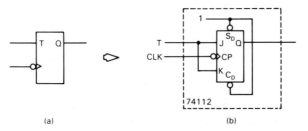

(a)   (b)

Fig. 5.6 Representations of a flip-flop. (a) T-type flip-flop; (b) 74112 used to implement a T flip-flop.

chips to be connected so that they are operating in a restricted way, and in such cases a reduced model is appropriate for testing purposes. A spare NOR or NAND gate, for example, is often used with its inputs joined together; it is then being used as an inverter, and should be treated as such.

To test the flip-flops in this circuit, therefore, we will simply check the basic functions of a clocked T flip-flop; with $T = 0$, the application of a clock pulse produces no change in the output (whichever state the flip-flop is in); with $T = 1$, the application of a clock pulse changes the state. There is nothing else that the device can do; four tests are required, represented by the four combinations of Q and T prior to the application of the clock pulse.

5.4.3 Initialization problems

In section 5.3.2, when discussing the testing of an isolated flip-flop, the point was made that it is impossible to know what state the device will enter when first switched on. The same is true of any sequential circuit, and so, since its behaviour depends not only on the inputs applied but also on its current state, it is essential that, before the test sequence begins, known conditions are established at every node in the circuit. These 'initial conditions' must be independent of the starting state of the circuit.

In the circuit of Fig. 5.5, these conditions are not easy to fulfil. In the absence of direct access to the flip-flop inputs, control over the state can be exercised only by setting up sensitive paths from the circuit inputs through the interconnecting gating. Some of the other inputs of these gates, however, are derived from flip-flop outputs, and so are functions of the state we are trying to control. A further difficulty of initialization derives from the fact that we are using T flip-flops. Because the only control possible is to 'change state' or 'don't change state', the flip-flops cannot be forced to specific values independent of the starting state.

One way of dealing with this problem could be simply to clock the system repeatedly until it reaches a particular state, say 000. Most testers have facilities to permit this to be done using a conditional instruction of the form

IF STATE ≠ 000 THEN CLK

There are, however, serious objections to this procedure:

  a It is not necessarily obvious whether any particular state will ever be reached. In the circuit of Fig. 5.5, for example, we have first to observe that with $A = 0$ the circuit will not change state at all. This is easy to verify once it has been pointed out, but less easy to deduce from the circuit diagram unaided. When $A = 1$, the circuit follows the state transition diagram shown in Fig. 5.7; the function of the circuit is simply a self-starting five-state counter whose normal operation is to circulate

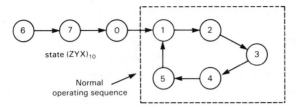

state (ZYX)$_{10}$

Normal
operating sequence

Fig. 5.7   State transition diagram for the circuit of Fig. 5.5.

round states 1, 2, 3, 4, 5. The remaining (redundant) states will never be entered except perhaps by chance at switch-on.

b Even if the chosen state is in the sequence under fault-free conditions, a faulty circuit may never reach it. Some kind of time-out facility would then be needed to avoid having the tester settle into an infinite loop.

c A circuit with many flip-flops, and hence many states, may give rise to a very long initialization procedure. A twenty-bit counter chain, for example, switching on in a state other than 00. . . . . .00 could then require over a million clock pulses to return it to all zeros. This waiting time may well be unacceptable, bearing in mind particularly that, on a complex board, the initialization procedure is typically repeated many times.

The conclusion to be drawn from all this is very simple: specific initialization facilities should be provided for test purposes even if they are not needed in normal operation. A simple provision in the present case could be as in Fig. 5.8; with CLR high (or disconnected), the circuit reverts to the form and function of Fig. 5.5; with CLR low, all flip-flops are simultaneously taken to the zero state irrespective of their current state.

5.4.4   Completing the test program

We are now in a position to generate a test program following the general strategy outlined in section 5.4.1. The circuit will first be initialized, assuming that the arrangement of Fig. 5.8 (or some equivalent provision) is implemented. We will then apply tests each of which is chosen by reference to

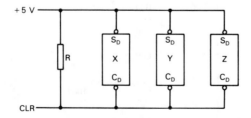

Fig. 5.8   Modification of the circuit of Fig. 5.5 to allow for initialization.

the current states of the flip-flops, and fault-cover is then checked for all elements in the circuit. The approach we will adopt to checking fault-cover will differ slightly from what has gone before; instead of making a fault-list we will instead form a list of required input conditions for each element. These conditions represent the functional tests for the flip-flops, and the structural tests for the gates. Hence the test conditions to be covered are:

a for each flip-flop, the four combinations of Q and T; these will be denoted by X_{00}, X_{01}, etc.;

b for each n-input gate, the $n + 1$ input combinations corresponding to the single-stuck fault-cover; these will not be denoted by $G1_{11}$, $G1_{10}$, etc., where the order of the suffixes represents the inputs reading from top to bottom in Fig. 5.5.

The criterion by which we judge whether the test conditions have been covered by a particular test is the same as with fault-cover. It is not enough that the input conditions are applied to the element; a wrong output from the element must be transmitted to a primary output where it can be observed.

The final question that needs to be considered is a test for the CLR line. If the circuit fails to initialize because of a fault on this line, the failure will, of course, be detected. On the other hand, if the circuit does initialize correctly, this does not necessarily show that the CLR line is fault-free, since it is quite

Table 5.4
Test set for circuit of Fig. 5.5

Test number	State ZYX	Circuit inputs CLR	A	Flip-flop inputs $T_Z T_Y T_X$	Next state $(ZYX)^+$	Test conditions covered
1	?	0	—	—	000	(Initialization)
2	000	1	1	001	001	
3	001	1	1	011	010	
4	010	0	—	—	000	CLR on Y
5	000	1	1	001	001	
6	001	1	1	011	010	
7	010	1	1	001	011	
8	011	1	1	111	100	
9	100	1	1	001	101	
10	101	0	—	—	000	CLR on X and Z
11	000	1	0	000	000	$X_{00}\ Y_{00}\ Z_{00}\ G1_{01}$
12	000	1	1	001	001	$X_{01}\ G1_{11}\ G2_{10}\ G5_{110}$
13	001	1	0	000	001	X_{10}
14	001	1	1	011	010	$X_{11}\ Y_{01}\ G2_{11}\ G3_{00}\ G4_{101}$
15	010	1	0	000	010	Y_{10}
16	010	1	1	001	011	$G4_{110}$
17	011	1	1	111	100	$Y_{11}\ Z_{01}\ G3_{10}\ G4_{111}$
18	100	1	0	000	100	$Z_{10}\ G5_{011}$
19	100	1	1	001	101	
20	101	1	0	000	101	$G4_{011}$
21	101	1	1	100	001	$Z_{11}\ G1_{10}\ G2_{01}\ G3_{01}\ G5_{11}$

possible for the switch-on state to be the same as the initialized state. The ideal test would be to set the circuit in the 111 state and then to activate CLR; this would check the whole function simultaneously. In this circuit, however, we cannot reach the 111 state, so we will have to test the CLR function in two stages, by initializing from 010 and then again from 101.

The whole program is shown in Table 5.4. It is of course assumed that, after setting the input values for each test, a clock pulse is applied to effect the state changes, except on lines 1, 4, and 10, where the use of the asynchronous inputs makes the use of the clock unnecessary (and, indeed, undesirable).

The first ten lines of this program constitute the sequence to test the CLR line; the dashes on lines 1, 4, and 10 indicate the 'don't care' conditions corresponding to the use of the asynchronous inputs to the flip-flops.

There is one test condition that is not covered by this program: $G5_{101}$. Since the inputs to G5 are X, $\overline{Y}$ and Z, this condition would require the circuit to be in state 111, which is one of the unattainable states. To see whether the loss of this test condition is significant, we can look again at the standard test set for a three-input NAND gate, as shown in Fig. 5.9.

There are two faults covered by the 101 test; $D/0$ is covered by other tests in the set, and $B/1$ is a fault that cannot be transmitted through the gate except when the inputs are 101. We can conclude, therefore, that, although there is an untestable fault, the presence of this fault would not affect the normal operation of the circuit.

It is perhaps worth pointing out that this is an example of an untestable fault that is not due to logical redundancy. The problem is caused by the existence of spare states; but in any sequential circuit this is almost inevitable, since the number of states required is determined by the particular application, whereas the number provided by the hardware has to be a power of two.

5.4.5 Problems of observability

Producing a test program for the circuit of Fig. 5.5 was a fairly painless process because we had direct access to the outputs of the flip-flops. Without this access, the problem can become very much more difficult with only a very modest increase in the complexity of the circuit. Suppose, for example, that the system requirement is for a timing waveform ϕ, as shown in Fig. 5.10. This could be obtained from the circuit of Fig. 5.5 by adding the output circuitry shown in Fig. 5.11. To test this modified circuit while being

Test	Faults covered
111	A/0 B/0 C/0 D/1
011	A/1 D/0
101	B/1 D/0
110	C/1 D/0

Fig. 5.9 Minimal test set and fault-coverage for three-input NAND gate.

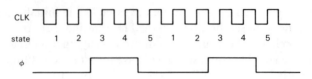

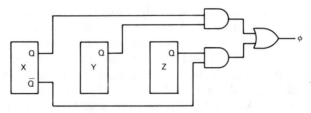

Fig. 5.10 Waveform to be derived from circuit of Fig. 5.5.

Fig. 5.11 Output logic to be added to circuit of Fig. 5.5 so as to provide the waveform of Fig. 5.10.

able to observe the flip-flop states only indirectly at the output of gate G8 immediately leads to difficulties: the sequence intended to check the CLR facility, developed in the first ten lines of Table 5.4, will no longer serve the purpose. Line 4 sought to test the CLR on Y; but with the new circuit we see that ϕ will be the same whether Y clears or not. It is, in fact, difficult to be certain about either X or Y because ϕ will always be low after a CLR even if either X or Y has failed to clear. The best we can do is to test for possible failures by observing a sequence of outputs. A suitable routine for this circuit could take the following form:

1 Activate CLR. ϕ should then be low; if it is not then the circuit is certainly faulty; but if ϕ is low we cannot say more than that the circuit must be in one of the states 0, 1, 2, or 5, assuming that the output circuitry is fault-free.

2 Clock the circuit three times. The fault-free behaviour is that ϕ remains low after the first two clock pulses, and goes high after the third. The same behaviour would be observed if the circuit had been in state 5 after step 1; the other two faulty states would be revealed by ϕ going high after one clock pulse (state 2) or two clock pulses (state 1). Correct behaviour therefore implies that the circuit is now in state 3: Z is low and X and Y are both high.

3 Activate clear. ϕ should go low, but we know from this only that at least one of X and Y has gone low: we have moved to one of states 0, 1, or 2.

4 Clock the circuit three times. This now distinguishes the three states (as in step 2) and can now be fairly said to have checked the CLR on X and Y.

5 Clock the circuit once more. ϕ should remain high, and the circuit is then in state 4: Z is high and X and Y are low.

6 Activate clear. ϕ should go low, to verify CLR on Z.

At the end of this sequence we can feel confident that the CLR facility has been fully checked. The possibility of a fault in the output circuitry (G6, G7 and G8) escaping detection in the sequence seems intuitively very unlikely, but should be investigated; a check through the sequence will show that any single-stuck fault in this part of the circuit will, in fact, cause an error at the output.

Turning to the main test program, we can start by considering the previous sequence (lines 11 to 21 in Table 5.4) and re-evaluating the fault-cover. On line 11, for example, we can see that the only test condition now covered is Z_{00}. As with the CLR routine, it may be necessary to introduce sequences of tests in order to complete the fault-cover.

It is evident that the reduction of observability brought about by this very simple block of output logic has made an enormous difference to the ease of test program generation; it may even have made some faults untestable. The difficulties will be compounded if the pathway between a flip-flop and a primary output contains not only combinational logic but also other flip-flops. The output logic of our waveform circuit, for example, was designed as shown in the Karnaugh map of Fig. 5.12(a); we require the output to be high when the circuit is in states 2 and 3, states 0, 6 and 7 being dont cares; this leads to the output logic

$$\phi = X.Y + \overline{X}.Z$$

We could, however, equally well have chosen to implement the output circuit as shown in Fig. 5. 12(b), giving

$$\phi = X.Y + \overline{X}.\overline{Y}$$

In this case, the performance of Z can be inferred only by the effect it produces on X and Y, as observed indirectly through the output logic at ϕ.

It is remarkable that we have here a circuit consisting of no more than three flip-flops and eight gates, which would fit into six chips leaving several gates to spare, but which is presenting a significant test-pattern generation problem. It is easy to see that, in a circuit of more realistic complexity, the cost of providing for automatic testing could easily become prohibitive. For a real product, which has to be produced at a commercially acceptable price, there must be a limit to the time and effort that can be devoted to the develop-

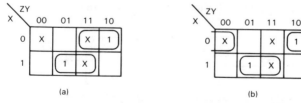

Fig. 5.12 Two ways of obtaining ϕ from XYZ. (a) $\phi = X.Y + \overline{X}.Z$; (b) $\phi = X.Y + \overline{X}.\overline{Y}$.

ment of a test program, so that the test engineer has to be prepared, if necessary, to accept a program that covers less than 100% of the faults in a circuit. Those that remain may be untestable, or they may simply be those that have defied the ingenuity of the programmer. Either way, the manufacturer is left hoping that these faults will not occur; especially he must hope the faults will not occur during maintenance or field-servicing when diagnosis would clearly be hampered by the inadequacy of the test program, and the cost of repair could become out of all proportion to the value of the circuit. It is for this reason that design modifications to improve testability are often accepted in industry as being economically justifiable even if they increase the initial hardware cost of the circuit.

5.5 COMBINATIONAL LOGIC WITH FEEDBACK

A major testing problem is posed by asynchronous circuits consisting of combinational logic with feedback. The simplest example of such a circuit is the basic R–S latch shown in Fig.5.13. This circuit is manageable (apart from needing care with timing, as discussed in section 5.2.2) because the function of the circuit is recognizable and it can, therefore, be tested functionally. If, however, we have a circuit such as that shown in Fig. 5.14, where feedback is applied globally around a block of combinational logic (which might, indeed, incorporate local feedback within itself) then the problem takes on new dimensions:

a If the circuit is drawn out clearly as in Fig. 5.14, the feedback path is clear; but if this circuit is embedded in a mass of logic, where the layout of the diagram may well be more closely related to the physical layout of the components than to the logical function of the circuit, then even the existence of the feedback path may be far from obvious.

b Having observed the feedback path, and hence deduced that the circuit is probably a sequential one, the test engineer may well find that the function of the circuit is not self evident. Few readers, for example, are likely to have guessed that in Fig. 5.14 the output Z can be stable in either state when $A = 1, B = 0, C = X, D = 1, E = 1$. If the function of the circuit cannot readily be deduced, the only test strategy open is a structural one with all the problems that entails.

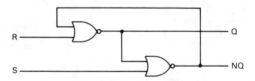

Fig. 5.13   The simplest asynchronous circuit: the RS latch.

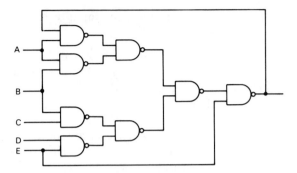

Fig. 5.14 Combinational logic with global feedback forming an asynchronous sequential circuit.

   c Initialization is a problem, as with all sequential circuits. In this particular case, analysis shows that

$$Z = 0 \text{ if } E(\overline{A} + \overline{B} + \overline{C}.\overline{D}) = 1$$
$$Z = 1 \text{ if } \overline{E} + A.B.C.\overline{D} = 1$$

Analysis, however, is an option that is not particularly easy even with this circuit, and is impractical with a more complex one. Finding a reliable initialization routine then can present a major obstacle to the test pattern generation task.

   d Having solved all the problems outlined above, and having produced a complete test program that reliably identifies a faulty circuit, we are then faced with the fault-location task. The presence of global feedback can make conventional probing techniques, as described in Chapter 4, ineffective; a fault within a feedback loop will cause faulty sequences all round the loop so that there is no element with correct inputs and faulty output (see section 4.5)

The problems posed by asynchronous logic, as outlined above, have a considerable bearing on approaches to design for testability, which will be discussed in Chapters 7 and 8.

SUMMARY
Chapter 5

The operation of a sequential circuit is substantially more difficult to describe than that of a combinational circuit. The basic component of sequential circuits, the flip-flop, is available in several different forms, and these different forms have significant differences in the way that they respond to changes of input. A general functional description of a flip-flop that will take accurate account of its temporal behaviour is difficult to devise, and the generation of a test sequence to cover all possibilities can present an intractable problem if the designer applies

Summary
Chapter 5
continued

logic signals to the clock input or to the asynchronous inputs.

Any sequential circuit, whether a single flip-flop or a system containing many flip-flops, must be initialized before testing can begin. The initialization sequence must be such that, at the end of the sequence, the circuit is guaranteed to be in a particular state (i.e. the logical value on every node is known) irrespective of the state from which the circuit starts. The circuit designer can make this task trivially easy if he provides access to the appropriate terminals; most straightforwardly, these would be the asynchronous inputs of all the flip-flops, but initialization through the data inputs can be equally effective. Failure to make this provision can result in a circuit for which initialization is very difficult, or even impossible.

The strategy employed for testing a sequential circuit is often a hybrid one, using functional testing for blocks whose functions are recognizable, and structurally based tests for the interconnecting logic. This strategy is often imposed on the test engineer because he has no knowledge of the overall function of the circuit. Having adopted this (or, for that matter, any other) strategy, the ease of implementation of the strategy depends crucially on the controllability and observability of the internal nodes. A design that improves these features of the circuit will use more silicon, but may prove to be cheaper in the long run.

Finally, the use of asynchronous logic, consisting of combinational circuitry with global feedback applied around it, presents a major testing problem, having an unclear function, hard-to-define operating characteristics, uncertain timing properties, and no obvious initialization facility. The use of such asynchronous circuits will always pose testing problems, so that modern design methodology increasingly seeks to discourage or forbid their use.

EXERCISES
Chapter 5

E5.1 The data sheet specifies set-up and hold times (t_s and t_h) for the 7473 (*JK* flip-flop) as shown in diagram (a), and for the 74LS73 as shown in diagram (b). Comment on the implications as regards the internal structure of the devices.

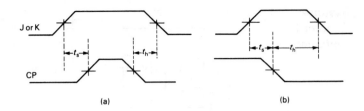

(a)   (b)

E5.2 A positive-edge-triggered D flip-flop is provided with an active-low asynchronous clear input, and has only its true output available.

Develop a test sequence for this device:

a on the basis of single stuck-pin faults;

b on a functional basis.

E5.3 The D notation introduced in Chapter 3 can be extended to a circuit containing stored-state devices. Establish the rules (analogous to Fig. 3.3) for a T flip-flop:

a to generate $D/\overline{D}$ at the output;

b to transmit input $D/\overline{D}$ to the output;

c to block transmission or produce fixed values at the output.

E5.4 The diagram shows a synchronous sequential circuit (the clock inputs are omitted for clarity) in which A, B, and C are primary inputs and Z is the primary output. There are no asynchronous inputs available on the flip-flops.

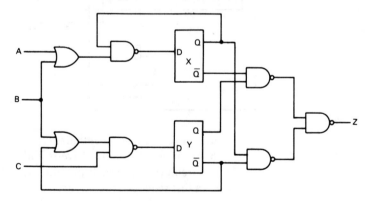

a How can the circuit be initialized?

b What strategy would you adopt for testing the circuit?

c Develop a test sequence for the circuit.

E5.5 The flip-flops in the circuit shown are driven by a

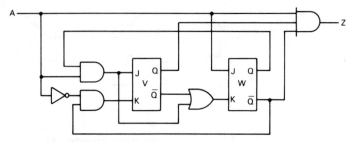

common clock. Find an initialization sequence for this circuit.

(Note: initialization means reaching a defined state irrespective of starting state.)

Assuming that the flip-flops do not have asynchronous inputs, suggest a modification to the circuit that would make initialization easier.

E5.6 The circuit below contains three *JK* flip-flops driven from a common clock. The only nodes accessible to the tester are the inputs *A* and *B*, and the output *Z*.

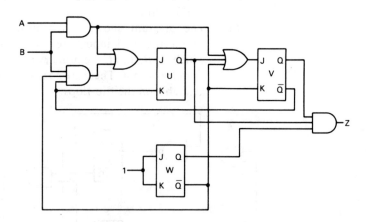

a can this circuit be initialized?

b If an asynchronous drive is added so that the circuit can be directly put into state 101, suggest a test sequence for the circuit.

6

TESTING MSI AND LSI DEVICES

6.1 COMBINATIONAL ELEMENTS

6.1.1 Gate-level descriptions

In building up a complex circuit, whether on a pcb or in a VLSI chip, the designer will not normally wish to be restricted to working with the simple basic elements (isolated gates and flip-flops) that have been considered in the previous chapters in this book. Although the basic elements will always need to be available, much of the design of a complex circuit is likely to make use of larger functional units such as multiplexers, decoders, code converters, arithmetic units and so on. Many such units are available to chip designers as macros within computer design systems, and a wide range of available MSI chips cater similarly for the needs of system designers.

One way of dealing with such elements is to work from the gate-level equivalents such as are specified for MSI components in the manufacturer's data sheets. For example, the SN74157 is a quad 2-1 multiplexer chip, whose equivalent circuit, as given in the data sheet, is shown in Fig. 6.1. A test program for this device can be derived using a structural approach based on the single-stuck fault-model. The circuit has 42 possible fault-sites (including fan-out branches) and hence 84 single-stuck faults. The test pattern generation task can be simplified by considering the circuit with just one multiplexer as shown in Fig. 6.2; each derived test can be extended to the full circuit by applying the same data signals to all four multiplexers and observing the responses at the four outputs. The minimal test set, covering all single-stuck faults, for the circuit of Fig. 6.2 is

$$ABCD/N = \{01X0/1;\ 01X1/0;\ 1X10/1;\ 1X11/0;\ 0010/0;\ 1100/0\}$$

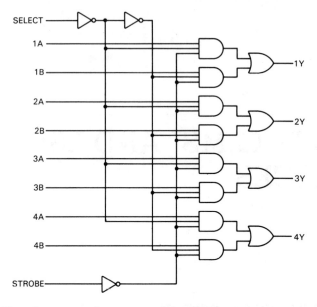

Fig. 6.1   Data sheet equivalent circuit of the 74157.

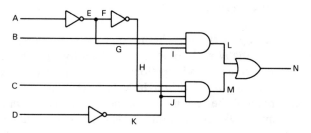

Fig. 6.2   One MUX from the 74157.

When this element is used as a component embedded within a larger design, however, the use of a gate-level description presents a number of difficulties to the test programmer.

6.1.2   Functional descriptions

The designer of a large and complex system has always to contend with a conceptual problem of deducing the effects produced throughout the system by changes occurring at any particular point. Although a detailed circuit diagram contains all the information needed to make these deductions, it is generally easier for a designer to work with larger functional units: a block diagram is much easier to understand than a detailed circuit diagram.

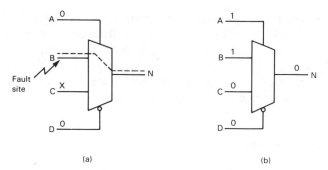

(a)   (b)

Fig. 6.3   Functional-level view of a MUX. (a) Path sensitization; (b) fault cover evaluation.

Consider, for example, the processes of establishing or assessing sensitive paths through a circuit containing a multiplexer. The circuit of Fig. 6.2 can be represented as shown in Fig. 6.3(a), and it is then obvious that if a fault-effect has been transmitted as far as node B, the conditions necessary to transmit it on to node N are $A = D = 0$, the value at C being immaterial. Equally, if a particular test establishes fault-free values at $ABCD$ of 1100, it is easily seen by inspection of Fig. 6.3(b) that single-stuck faults on A and C will be transmitted, while those on B and D will not.

The same deduction could, of course, be obtained from the circuit of Fig. 6.2, but, without making use of the knowledge of the overall function, this requires the application of the full analytical techniques developed in Chapter 2.

6.1.3   Diagnostic needs

The smallest replaceable unit on a pcb is a chip. Hence, diagnosis to an internal node or to an individual embedded gate is not required. There is, therefore, a case to be made for generating a test set using either a structural approach based on pin-faults or a functional approach.

A minimal test set for the multiplexer of Fig. 6.2 using pin-faults only (that is, single-stuck faults on nodes A,B,C,D and N) consists of five tests, compared with the six needed to cover all single-stuck faults. Analysis shows that the tests $1X11$ and $01X1$, both of which are needed to give full fault-cover, appear as alternatives in the pin-fault-based test set. The result is that either $F/0$ or $G/0$ will remain uncovered. In this particular case this result might be considered acceptable (particularly in the light of the points to be considered in section 6.1.4), but not all circuits are as well-behaved as the multiplexer.

A functional test program could take the form of an exhaustive test set, which would cover all irredundant faults without the need for any TPG effort, but which would generally use far more test patterns than the minimal test set. Alternatively, an intuitive interpretation of the functional specification of the device can be invoked, exercising all the data and control inputs. The fault-cover of such a test set will be generally unpredictable, but is likely to be at least as good as the pin-fault test set. A major disadvantage of the functional approach is the difficult of automating it; here the inability to generate fault-cover figures is a major handicap.

6.1.4 Available data

The most important reason for questioning the use of the gate-level equivalent circuit as the basis for deriving test sequences for commercially available chips is that accurate knowledge of the internal structure is not generally available. The data sheet is not reliable in this respect: as pointed out in section 5.3.1, the circuit diagram supplied is usually described as a 'functional block diagram', and there is no way of knowing how closely this corresponds to the actual circuit on the chip. If the data sheet circuit is used as the basis of our TPG, therefore, we may well be deriving tests to detect faults on internal nodes that do not, in fact, exist. Equally, there may well be other nodes in the physical circuit that are not represented in the data sheet circuit. Fig. 6.4, for example, shows the data sheet diagram of the SN74180 (parity generator). One cannot help wondering whether the manufacturer actually uses five different types of gate; in particular, it is likely that an exclusive-OR 'gate' is actually composed of several primitive gates connected together. If this is so, there will be additional internal nodes not represented in the circuit of Fig. 6.4, and some of these nodes will have fan-out, producing particularly difficult-to-test faults.

All these considerations must lead us to question the validity of trying to rely exclusively on gate-level equivalent circuits for TPG purposes when dealing with MSI components. The use of a functional approach, perhaps

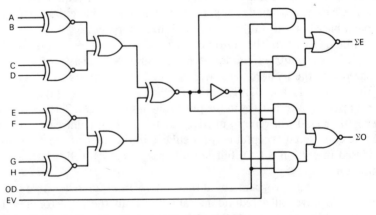

Fig. 6.4   Data sheet equivalent circuit of the 74180.

Table 6.1
Functional specification of the 74180

Inputs			Outputs	
Data (A – H)	OD	EV	ΣE	ΣO
Even no. of '1'	0	1	1	0
Odd no. of '1'	0	1	0	1
Even no. of '1'	1	0	0	1
Odd no. of '1'	1	0	1	0
X	1	1	0	0
X	0	0	1	1

Table 6.2
Possible functional test for the 74180

	Data inputs								Control inputs		Fault-free outputs	
	A	B	C	D	E	F	G	H	OD	EV	ΣE	ΣO
1	0	0	0	0	0	0	0	0	1	1	0	0
2	0	0	0	0	0	0	0	0	0	0	1	1
3	0	0	0	0	0	0	0	0	1	0	0	1
4	0	0	0	0	0	0	0	0	0	1	1	0
5	1	0	0	0	0	0	0	0	0	1	0	1
6	1	1	0	0	0	0	0	0	0	1	1	0
7	1	1	1	0	0	0	0	0	0	1	0	1
8	1	1	1	1	0	0	0	0	0	1	1	0
9	1	1	1	1	1	0	0	0	1	0	1	0
10	1	1	1	1	1	1	0	0	1	0	0	1
11	1	1	1	1	1	1	1	0	1	0	1	0
12	1	1	1	1	1	1	1	1	1	0	0	1

followed by a check that all single-stuck pin-faults are covered, is a strategy that can more easily be defended. For the circuit of Fig. 6.4, for example, whose functional specification is outlined in Table 6.1, a functionally-based test sequence might be as in Table 6.2.

The derivation of Table 6.2 is straightforward; the first two lines check the two disabled states of the circuit (corresponding to the last two lines of Table 6.1); lines 3 and 4 complete the check on the control modes; lines 5–12 check the normal operation. This test set (which was formulated intuitively and is certainly not unique) covers all single-stuck pin-faults, and, in fact, also covers single-stuck faults on all nodes shown in Fig. 6.4, including the fan-out branches. It is interesting to observe that a test set consisting only of tests 4 and 12 from Table 6.2 would cover all single-stuck pin faults but would miss a significant number of faults on the internal nodes. This problem is always likely to be troublesome in any circuit performing arithmetic functions (the parity generator is essentially counting the number of 1 inputs). In such circuits, the XOR function plays a prominent part, and it is a property of the XOR gate that changing both inputs makes no change to the output. The functional test represented by Table 6.2 was based on the intuitive feeling that each data input should be shown to have an individual effect on the outputs.

When it comes to dealing with larger functional elements, such as microprocessors and their support chips, the test programmer is given no choice in the matter: the manufacturer does not normally offer anything remotely like a gate-level equivalent circuit. A functionally-based test generation procedure is the only option available in this case.

6.2 PROGRAMMABLE LOGIC ARRAYS

6.2.1 Principles of operation

The high cost of production of the complete mask set for an ic means that a full custom chip cannot be manufactured economically unless an extremely large market (many millions of units) exists for that chip. One way round this problem when designing a special purpose system with a relatively low volume market is to use standard SSI and MSI elements on a custom-designed pcb to form a so-called **random logic** circuit. Such a design benefits from the cost advantages of using mass-produced components, but tends to be bulky and to consume a lot of power. An alternative approach is to use much more complex devices that can address a mass market because they are capable of performing any one of a large range of functions, being adapted to a particular purpose by being programmed. The outstanding example of such a device is the microprocessor, which will be considered in section 6.4. A somewhat simpler member of this class of elements is the **programmable logic array** (PLA), which is particularly important as a component in VLSI designs, where its regular structure makes it especially suitable.

Figure 6.5 shows a circuit that generates three independent functions of three variables, using a two-level realization consisting of a layer of AND gates

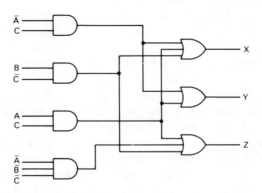

Fig. 6.5 Two-level multiple-output combinational logic circuit.

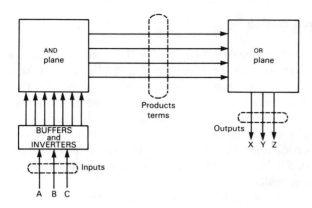

Fig. 6.6   General structure of a PLA.

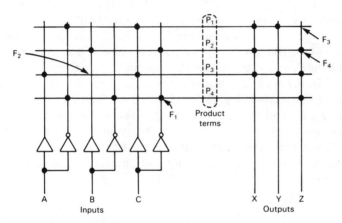

Fig. 6.7   PLA programmed to implement the same functions as the circuit of Fig. 6.5.

followed by a layer of OR gates. By choosing which particular inputs are connected to the AND gates a particular set of product terms will be defined; the outputs are then formed by choosing some of the product terms for connection to the OR gates. One form of PLA consists of a structure of the form of Fig. 6.6, with, initially, no connections to either the AND or the OR gates. Programming the PLA consists of making the connections necessary to realize the functions required by the particular user. Figure 6.7 is a diagrammatic representation of a PLA programmed to realize the functions shown in Fig. 6.5; the points at which the horizontal and vertical lines cross are known as **cross-points**, and are the points at which connections are or are not made when the PLA is programmed.

6.2.2 Fault-models

While the details of the construction of a PLA need not concern us here, it is important to realize that the diagram of Fig. 6.7 bears a close resemblance to the physical structure. Each cross-point connection takes the form of a transistor connected between the two lines, and most of the elements of this transistor exist at every cross-point as part of the basic structure of the device. Hence the physical difference between a cross-point connection and no cross-point connection is small, and it is easy to envisage defects whose effects can be represented by either additional or missing connections. These are known as **cross-point faults**.

In Fig. 6.7 four possible sites for cross-point faults are indicated (F_1-F_4). The effects of these particular faults are most easily appreciated by reference to the Karnaugh maps describing the operation of the circuit. Fig. 6.8(a) shows the four product terms, $P_1 - P_4$, in the fault-free circuit. The effects of faults in the AND plane are shown in Fig. 6.8(b); fault F_1 is a missing connection between the $\overline{C}$ input and the P_4 AND gate, causing P_4 to **grow** from $\overline{A}.\overline{B}.\overline{C}$ to $\overline{A}.\overline{B}$; conversely, F_2 is an additional connection between the B input and the P_3 AND gate, causing P_3 to **shrink** from $A.C$ to $A.B.C$. Faults F_3 and F_4 in Fig. 6.7 are in the OR plane; F_3 is an additional connection, while F_4 is a missing connection. Figure 6.8(c) shows the fault-free function Z consisting of P_2, P_3, and P_4; F_3 causes the **appearance** of P_1, while F_4 causes the **disappearance** of P_2; the resulting output functions are shown in Fig. 6.8(d) and (e). These four effects constitute the most commonly used fault-model for PLAs. A single fault assumption is usually made; this is necessary (as with

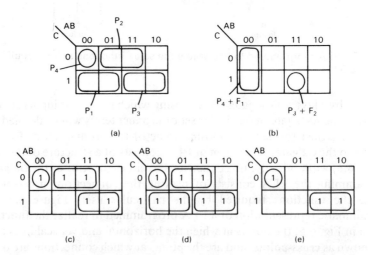

Fig. 6.8 Effects of cross-point faults in the PLA of Fig. 6.7. (a) The four product terms; (b) effects of growth (F_1) and shrinkage (F_2) faults on the product terms; (c) Fault-free output Z; (d) effect of appearance fault (F_3) on output Z; (e) effect of disappearance fault (F_4) on output Z.

the stuck-fault model) to keep the fault-list within bounds, but in fact it has been shown that most multiple faults will in any case be detected by a test set that covers all single faults.

The fault-list for the circuit of Fig. 6.7, therefore, will consist of the following:

Growth	$P_1 \rightarrow \bar{A}, C$
	$P_2 \rightarrow B, \bar{C}$
	$P_3 \rightarrow A, C$
	$P_4 \rightarrow \bar{A}.\bar{B}, \bar{A}.\bar{C}, \bar{B}.\bar{C}$
Shrinkage	$P_1 \rightarrow \bar{A}.\bar{B}.C, \bar{A}.B.C.$
	$P_2 \rightarrow \bar{A}.B.\bar{C}, A.B.\bar{C}$
	$P_3 \rightarrow A.B.C, A.\bar{B}.C$
	$P_4 \rightarrow 0$
Appearance	P_1, P_2, P_3, P_4
Disappearance	P_1, P_2, P_3, P_4

Notice that P_4 can shrink to nothing because of an additional connection in the AND plane, resulting in both a variable and its inverse being connected. It can also disappear from the Z output because of a missing connection in the OR plane. Since this is the only output affected by P_4 it is clear that these two faults are indistinguishable at the output although they arise from totally different defects. Rules for fault collapsing (both equivalence and dominance) have been formulated for PLAs and would allow the fault-list above to be significantly reduced.

6.2.3 Test generation

Having established the fault list, the process of test generation follows essentially the same lines as before; conditions must be set so that there will be a difference between the fault-free circuit and the circuit containing the particular fault being considered; further conditions must be chosen so that this difference will be transmitted to an output. Consider, for example, growth faults on P_4 in the circuit of Fig. 6.7. We can see first of all that three such faults are possible: $\bar{A}.\bar{B}.\bar{C}$ can grow to $\bar{A}.\bar{B}$ (as illustrated in Fig. 6.8(b)), $\bar{A}.\bar{C}$ or $\bar{B}.\bar{C}$. By supplying the circuit with inputs 001 we ensure that for the fault-free circuit P_4 will be 0, while if the first of the P_4 growth faults is present, P_4 will be 1. We next need to choose an output for observation: in this case Z is the only output at which a difference between faulty and fault-free circuits would be observable.

Assessment of fault coverage for a PLA also follows essentially the same lines as for other circuits; the principle is simply to identify all faults on the list that would cause at least one of the outputs to have a value different from its fault-free one. The input vector 001, for example, will cover the following faults:

Growth	$P_4 \rightarrow \overline{A}, \overline{B}$	(Z = 0 → 1)
	$P_3 \rightarrow C$	(Z = 0 → 1)
Shrinkage	$P_1 \rightarrow \overline{A}.B.C$	(X = 1 → 0, Y = 1 → 0)
Appearance	P_1	(Z = 0 → 1)
Disappearance	P_1	(X = 1 → 0, Y = 1 → 0)

The number of product terms in a PLA is always much smaller than the number of minterms, relying on the fact that, in virtually all combinational functions required in practice, minterms can be combined to form larger terms. The circuit in Fig. 6.5, for example, generates three functions of three variables using four product terms, only one of which is a minterm. The fact that this circuit has multiple outputs, in common with most PLAs, influences the choice of product terms, with the result that the set chosen will not necessarily form a minimal set for any of the output functions; moreover, the different product terms may overlap wholly or partly.

These features can affect the testability of the circuit; undetectable faults can be present even without logical redundancy. In the circuit of Fig. 6.7, for example, the growth fault $P_1 \rightarrow C$ is undetectable since the only outputs containing P_1 also contain the grown term (AC). From a functional point of view this would not matter, since the PLA with this fault in it is functionally indistinguishable from the fault-free one, but a count of fault-cover, as might be used to direct the operation of an ATPG system, would be misleading.

6.3 READ ONLY MEMORY (ROM)

Although, from the user's point of view, a ROM is a memory device which when programmed will retain a fixed array of data, it is in many ways much more like a combinational circuit in which each output can be any arbitrary function of the address inputs. The first stage of a ROM is the address decoder, which can be thought of as the AND plane of a PLA with the important difference that the product terms are neither selected nor restricted: every minterm is generated. The process of programming the ROM consists of specifying the outputs (one or more bits) for each minterm, and this is implemented essentially by connecting the corresponding product term (decoder output) to each ROM output wherever a 1 is wanted, and leaving the connection open where a 0 is wanted. The actual mechanism by which this is achieved depends on the kind of ROM and the technology in which it is built. Fig. 6.9 shows a typical organization for a field programmable ROM (PROM) with n input (address) lines and 4 output lines. The decoder yields 2^n lines, each of which serves to identify one location in the memory defined by one minterm of the address variables. Each bit of

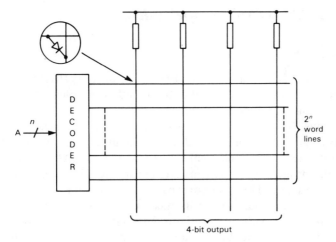

Fig. 6.9   Organization of a PROM.

the output is initially connected through diodes to all 2^n decoder lines (so that the output would be 1111 in response to every address); this output circuitry constitutes a diode-resistor-logic OR gate. If the diode is intact, the corresponding minterm is part of the output function, but if the diode is open circuit the minterm is absent from the output function. The process of programming the PROM then consists simply of destroying the diodes in all those bit positions in which a 0 is required.

Whatever may be the detailed mechanism whereby the ROM is built and programmed, it is clear that the only data available to the test programmer will be the truth table defining the required contents of the ROM. It is also clear that normally there will not be any identifiable functional relationship between the outputs and the address inputs. There is, in short, only one satisfactory way of testing a ROM: the whole truth table needs to be verified.

The obvious way to check a truth table is simply to step through the address space in any convenient order (not necessarily the binary sequence: in some cases an LFSR sequence may be more convenient – see section 9.2.2) and to compare the output at each address with the known fault-free output, which would be stored as part of the test program. Both the data storage and the test run time can be reduced by making use of data compression; the CRC signature (see section 4.2.2) is particularly suitable for this purpose, but other forms of data compression (checksums) can also be used.

It is worth noting in passing that in order to apply these test procedures it will be necessary for the tester to have access to both the address and the data lines for the ROM. This requirement is common to many large functional blocks (ROM, RAM, PLA, etc.) that are used as components in systems, and can make for difficulties in interfacing to the tester (see section 7.7.1). In particular there can be difficulties with a VLSI chip in which these functional blocks are embedded deep within the circuit.

6.4 RANDOM ACCESS MEMORY (RAM)

6.4.1 Structure

The structure of the random access read/write memory (RAM) is necessarily rather more complex than that of a ROM, and the problem of defining a test program that is both comprehensive and economical is a correspondingly difficult one. In what follows, we will consider for simplicity a RAM organized as single bits (a common, although not universal, organization).

The main features of the architecture of a typical RAM are illustrated in Fig. 6.10. The memory cells themselves are arranged in a rectangular matrix, and they are accessed by means of separate row and column decoders. Transfer of data into and out of the cells makes use of sense amplifiers, which are actually flip-flops with differential inputs and outputs, and the process is controlled by WE (Write Enable) and one or more CE (Chip Enable) inputs.

There are two main types of RAM, with important differences in the way they operate and hence in the ways in which they can fail. The first is the static RAM, in which the memory cells are flip-flops; these cells will retain their data indefinitely provided only that power remains applied to the chip. The dynamic RAM relies for data storage on charge stored on a capacitor; this results in a cell that is much smaller than a static RAM cell, which permits a much higher packing density. The penalty attached to this size advantage is simply that stored charge inevitably leaks away, so that if data is to be retained in the memory it must be refreshed at intervals (typically about

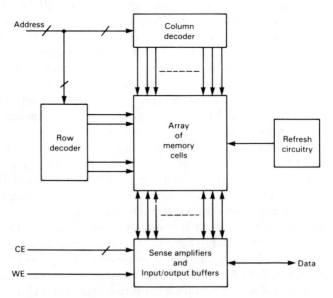

Fig. 6.10 Organization of a RAM. ('Refresh circuitry' required only for a dynamic RAM.)

2 ms). The circuitry necessary to provide refresh is often provided on the chip, in which case the process is transparent to the user.

Although the memory cells are arranged in a rectangular matrix and they are accessed by a pair of signals that can be regarded as co-ordinates, there is no guarantee that cells whose addresses are numerically adjacent are physically adjacent. There are two reasons for this:

a In laying out a decoder circuit on a chip, it turns out that from a processing point of view the most economical design, in terms of chip area and, particularly, the number of crossovers, has its outputs in a sequence that does not correspond to the normal binary count. In normal use, of course, this does not affect the user, who is only concerned with using a particular address to enter and retrieve data, and does not care where on the chip the cell is located. For those who do care (and we will see shortly that this includes the test engineer) the relationship between numerical and physical addresses is given by the manufacturer, so that a translation routine can be incorporated when needed; this process is known as **topological descrambling**.

b Every manufacturer of ics is engaged in a continual struggle to increase the yield of his process. The manufacturer of memory chips is in a slightly better position than most because of the nature of his product. Rather than being a single system, the memory chip largely consists of separate non-interacting elements, so that the possibility exists of including spare cells on the chip and using these to replace defective cells in the body of the array. In practice this is done by having spare rows and columns which are brought into play by a system of **discretionary wiring** consisting usually of fusible links. The resulting pattern of active cells is peculiar to the individual chip, and cannot therefore be descrambled. However, the number of such scrambled cells is only a small fraction of the whole array.

It may be thought that the simple and well-defined function of a memory chip would make structural considerations unnecessary; a functional test can surely be proposed, based on the requirement that every cell must be capable of storing both a 0 and a 1, and of returning the data when it is read. This functional description could be tested by filling the memory with 0 and reading every cell, then filling with 1 and again reading every cell. However, it is easy to demonstrate that this test is inadequate. If we consider the structure of the device, as illustrated in Fig. 6.10, it is clear that successful operation will involve all the circuitry used to access the cell as well as the cell itself. One effect that could be brought about by a classical stuck fault in the decoder might be that some cells are never accessed while others respond to more than one address. The limit would be a circuit in which every address accesses the same cell; this circuit would still pass the test outlined above!

6.4.2 Fault mechanisms

As may be expected with such a complex and tightly packed circuit, there are many ways in which a RAM can fail. The list that follows is certainly not exhaustive, but will serve to illustrate not only the variety of defects that can be encountered, but also, more importantly, the fault-effects that will demand special test procedures to detect them.

a Defects in metallization (either too much or too little) can produce open or short circuit. Other processing faults can result in breakdown in diodes or transistors. These are the kinds of defect that afflict all ics, and their effects are usually well modelled by bridging and stuck-at faults on the cells or other nodes in the circuit.

b Defects in the decoder, which could be modelled locally as stuck-at faults, produce externally functional failures in the addressing system. These can take two forms:

i some cells effectively have two (or more) addresses while other cells cannot be accessed;

ii some addresses access two (or more) cells simultaneously.

Defects of these first two types produce major and permanent deviations from correct behaviour, and as such are relatively easy to cater for. The next group of defects is more difficult because they produce degradation of performance rather than outright failure:

c Charge storage in the driving circuits, and particularly in the long address lines that are inevitable in a structure of this kind, can result in an increase in the access time required.

d The use of the same data line for input and output can bring problems because of the larger voltage swings that are involved in writing. If the time taken to recover after a write cycle becomes excessive, the circuit can fail, but this failure will normally become manifest only if the write cycle is followed immediately by a read cycle. Whether or not failure occurs may also depend on the particular patterns of data involved.

e The sense amplifiers can also be affected by unwanted charge storage. This can result in the amplifier becoming biased; after reading a long succession if 1s, a 0 may be read as a 1. The same amplifier will work perfectly when reading an alternating sequence of 1s and 0s.

f In a dynamic RAM, the leakage of stored charged can be such that data is not retained reliably for the specified refresh period. Since the data is automatically refreshed by the reading process, this fault can easily be overlooked unless it is specifically tested for.

Perhaps the worst problem presented by these types of defect is the fact that whether or not the fault-effect appears depends on the particular sequence of activity to which the device is subjected. The device might, indeed, work perfectly correctly most of the time, but produce occasional and apparently random errors. Such errors are known as **soft errors**; they have the effect of being apparently intermittent and unreproducible, in contrast to hard errors, typified by open or short circuit effects. This same feature characterizes the next type of defect, which has proved to be one of the most important of all:

g For reasons that are not entirely clear, but which probably involve a combination of marginal performance and parasitic coupling effects, it turns out that activity in one cell can affect data in another cell, the effects being strongly dependant on the pattern of data stored. A typical fault-effect of this kind can be described as:

If cell j is in state 1, a 0 $\rightarrow$ 1 transition in cell i will change the state of cell j.

These effects are known as **pattern-sensitive faults**.

Finally there are ways in which data can be corrupted in a truly random manner due to effects external to the circuit. Although these may be regarded as 'Acts to God', the sensitivity of the circuit to these external effects is certainly affected by various features of the circuit design.

h If an alpha-particle impinges on the active area of the chip, a packet of charge is injected into the circuit, and this can have the effect of corrupting the stored data. When this effect was first observed it posed a serious problem, the cause of which was tracked down to traces of radioactive elements in the substrate and packaging. Reasonably effective countermeasures have since been taken, but there are still some fears that larger memories (with smaller geometry) may be susceptible to this form of soft error.

i As with all circuits, transients and noise on power and data lines can be transmitted into the circuit, and can therefore corrupt data; this can be a particular hazard in a RAM because of the very small voltage differences that separate the two logic states.

6.4.3 Typical test procedures

From the daunting array of fault mechanisms discussed in the previous section, it will be evident that we cannot expect to find a single test procedure that will cover all possible faults. The main concern has been to cover the pattern sensitive faults, since a comprehensive test for these will cover the hard faults such as stuck cells and decoder failures. The test strategies described below are just a small selection of those that have been proposed in the literature; there is still no universal agreement in the testing community as to the ideal way of testing large memories.

a GALPAT
There have been many variants of the 'galloping ones' (or 'galloping zeros') test schemes, all known as GALPAT. One typical procedure for

a memory with N cells can be described by the algorithm below. One of the most important parameters of any memory test procedure is the number of test cycles needed to apply it; this is easily assessed by counting the number of read and write operations. This number is indicated for each step in the algorithm.

1 Reset all cells to 0.   [N cycles]
2 For i = 0 to N-1
    Read cell (i)
    Write 1 into cell (i)
    Read cells (0) to (i-1) and (i + 1) to (N – 1)   [$N(N + 1)$ cycles]
3 For i = N-1 to 0
    Read cell (i)
    Write 0 into cell (i)
    Read cells (N-1) to (i + 1) and (i-1) to (0)   [$N(N + 1)$ cycles]

This procedure checks interactions between each cell and each other cell, although it does not check all possible such interactions. Nevertheless, the cost of this program is considerable; the total number of cycles is $N(2N + 3)$, which applied to a 256K memory with a 450 ns cycle time would take over 17 hours to complete! All members of this class of test patterns, characterized by a test time proportional to N^2, are unusable for large memory chips.

b Nearest Neighbour Disturb (NND)

The cost of GALPAT lies in the check on interactions between each cell and every other cell. It seems plausible to suggest that if there is to be interaction between cells, it will be between cells that are physically adjacent. This is the basis of the NND test. With the rectangular grid array in which memory cells are arranged, each cell has eight neighbours (or they could be considered to have only four neighbours if the diagonal ones are excluded). By using the GALPAT procedure above, but reading only the eight neighbours, lines 2 and 3 of the procedure require only $8N$ cycles each, and the whole procedure requires $17N$. A 256 K memory can therefore be tested using this version of NDD in 2 seconds.

It should be noticed that the NND algorithm makes sense only if physically adjacent cells can be identified. Hence it is necessary to apply topological descrambling in implementing the algorithm, and to accept that discretionary wiring, if it has been used, will invalidate some of the comparisons. This, however, is a small price to pay for the economy of program size.

c MARCH

One of the earliest procedures to be suggested (and one that also has several variants) is the MARCH pattern. It is described by the following algorithm.

1 Reset all cells to 0   [N cycles]
2 For i = 0 to N-1
    Read cell (i)
    Write 1 into cell (i)   [$2N$ cycles]
3 For i = N-1 to 0
    Read cell (i)
    Write 0 into cell (i)   [$2N$ cycles]
4 Set all cells to 1   [N cycles]
5 For i = 0 to N-1
    Read cell (i)
    Write 0 into cell (i)   [$2N$ cycles]
6 For i = 0 to N-1
    Read cell (i)
    Write 1 into cell (i)   [$2N$ cycles]

For a total of 10N cycles, MARCH will cover most decoder failures (although some multiple addressing failures can be missed), but the coverage of pattern sensitive faults is low.

d Abraham's algorithm

By formulating a fault-list consisting of stuck cells together with pattern sensitivities of the kind described in section 6.1.2.(g), a procedure has been derived to cover all these faults in 30N cycles. It takes the form of an extended MARCH procedure.

1 Reset all cells to 0   [N cycles]
2 For i = 0 to N-1
    Read cell (i)
    Write 1 into cell (i)   [$2N$ cycles]
3 For i = N-1 to 0
    Read cell (i)   [N cycles]
4 For i = 0 to N-1
    Read cell (i)
    Write 0 into cell (i)   [$2N$ cycles]
5 For i = N-1 to 0
    Read cell (i)   [N cycles]
6 For i = N-1 to 0
    Read cell (i)
    Write 1 into cell (i)   [$2N$ cycles]
7 For i = 0 to N-1
    Read cell (i)   [N cycles]
8 For i = N-1 to 0
    Read cell (i)
    Write 0 into cell (i)   [$2N$ cycles]
9 For i = 0 to N-1
    Read cell (i)   [N cycles]

10 For $i = 0$ to N-1
    Read cell (i)
    Write 1 into cell (i)
    Write 0 into cell (i)   [3N cycles]
11 For $i = N$-1 to 0
    Read cell (i)   [N cycles]
12 For $i = N$-1 to 0
    Read cell (i)
    Write 1 into cell (i)
    Write 0 into cell (i)   [3N cycles]
13 For $i = 0$ to N-1
    Read cell (i)   [N cycles]
14 Set all cells to 1   [N cycles]
15 For $i = 0$ to N-1
    Read cell (i)
    Write 0 into cell (i)
    Write 1 into cell (i)   [3N cycles]
16 For $i - N$-1 to 0
    Read cell (i)   [N cycles]
17 For $i = N$-1 to 0
    Read cell (i)
    Write 0 into cell (i)
    Write 1 into cell (i)   [3N cycles]
18 For $i = 0$ to N-1
    Read cell (i)   [N cycles]

It is not easy to assess these test procedures, or the many others that have been proposed. Certainly we cannot afford, for the size of memories that are now being produced, test programs of length proportional to N^2, but beyond that it is difficult to make any general judgement.

The various routines could be assessed on the basis of fault-cover, but the value of such an assessment depends, as with all test programs, on the validity of the fault-models. There is certainly more work to be done in the research laboratories before this problem will be finally solved.

6.5 MICROPROCESSORS

6.5.1 The general problem

The microprocessor represents perhaps the most complex single unit being produced today, and it presents the test engineer with all the problems that have been emphasized throughout this book:

a access to the circuit is very limited, being largely confined to the address and data busses;

b there is little or no gate-level information available, and even block diagram-level information is generally sparse;

c a general statement of the function of the circuit is very difficult to formulate.

As with other LSI elements, we cannot expect classical fault-models to be very useful as a basis for formulating test programs for microprocessors; considerable effort has been expended (and is still being expended) on developing appropriate fault-models and usable functional descriptions. Some of the approaches that have been suggested are outlined below.

6.5.2 Architecture-based approach

A 'programming model' of the 6800 microprocessor as given by the manufacturer is shown in Fig. 6.11. An architectural approach to testing this device consists of attempting to exercise each of the elements of the model. A possible strategy could therefore be based on the operational requirements of the elements:

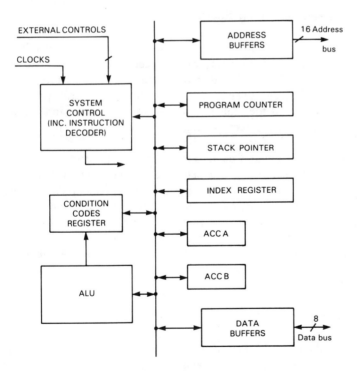

Fig. 6.11   Programming model of an MC6800 microprocessor.

1 The program counter must be able to increment throughout the address space, and to be set directly to any address.

2 Both the stack pointer and index register are required to increment and decrement throughout the range, and to transfer arbitrary data to and from the data bus. Each should also be able to transfer data to the other.

3 The two accumulators have the same requirements as the stack pointer and index register, except that they operate on operands of eight bits rather than sixteen.

4 The ALU is the most difficult block to test. It is required to perform all the data manipulations allowed in the instruction set, but the problem is to choose the operands so that correct operation is reliably demonstrated.

5 The condition code register consists of independent bits that should be controlled by the results of ALU operations, and in addition it should be possible to set and reset some of the bits separately.

6 The system control block will have been involved in the tests of all the other blocks, but in addition it will be necessary to verify the actions of the various external control lines.

6.5.3 Instruction set verification

The nearest approach to a complete functional description of a microprocessor is the instruction set, which specifies the response to all valid instructions, and this can be taken as the basis for a test strategy. In many respects this approach will be very similar to the architectural approach, since our interpretation of the fault-free operation of a register in the microprocessor will clearly have been strongly influenced by our knowledge of the instructions that are in the repertoire.

A purely functional approach to testing a microprocessor, whether based on the architecture or on the instruction set, suffers from the two difficulties common to all functionally based tests: the choice of operands has to be made largely intuitively, and there is no easy way to assess the comprehensiveness of the test set. Because of this, there have been a number of attempts to formulate a model of a microprocessor that can be used as a basis for structural TPG. The most impressive such attempt to have been made so far is the S-graph.

6.5.4 S-graph

A functional model of a microprocessor can be constructed using only information about its instruction set and the functions performed. The model takes the form of a system graph (the **S-graph**), in which each register is represented by a node. Further nodes represent the input and output. The flow of data among the registers, and to and from the input/output devices, is represented by paths, or directed edges, connecting the nodes, and each instruction is represented by a set of paths, showing the data flow involved in executing that instruction. The architecture of the microprocessor is represented as five modules which are defined on a functional basis, and which can, importantly, be specified without any detailed knowledge of the implementation details.

Faults are specified with respect to the functional modules, and, with the aid of the S-graph, test patterns can be generated automatically to cover these faults. The modules, and their associated fault-effects, are as follows:

1 Register decoding – when attempting to access a register, either no register is accessed or multiple registers are accessed.
2 Instruction decoding and control – when attempting to execute an instruction, either no instruction is executed, or the wrong instruction is executed; in the latter case, this can either be instead of or in addition to the required instruction.
3 Data storage – any cell of any register can be stuck at 0 or 1.
4 Data transfer – in a transfer bus, any wire can be stuck at 0 or 1, and any two wires can be bridged.
5 Data manipulation – this is largely the ALU data transformation function; it is tested by 'classical' methods.

It is particularly interesting to observe that when this approach was applied to a commercial 8-bit microprocessor for which a gate-level description was available, items 1 to 4 above were responsible for a test program of about 1 K instructions which covered about 90% of single-stuck faults. Item 5 was responsible for a further 8 K instructions, which covered a further 6% of single-stuck faults.

The problem of finding an adequate test set for a device of this complexity is not a trivial one, and the ideal solution has probably not yet been found. A number of recent microprocessors have been manufactured with built-in self-test programs incorporated; it may well be that this is the only way that such devices can realistically be tested.

SUMMARY
Chapter 6
The problems of testing the more complex functional units represented by LSI circuits can be seen as stemming from the inadequacy of the fault-models used for simple SSI circuits. The use of single-stuck and bridging faults at the nodes in the gate-level circuit diagram is of doubtful validity because in practice the test programmer does not have access to an accurate gate-level description. Pin faults appear attractive, but provide in many cases a seriously deficient fault-cover. The test program must be

Summary
Chapter 6
continued

based on a knowledge of the function of the circuit, together with an appreciation of the effects that can plausibly be expected due to the nature of the device and its construction.

The PLA, although seemingly just a two-level combinational circuit, will actually be prone to defects that cannot be represented as stuck-faults since they bring about radical changes in the logic function realized. The appropriate fault-model is based on cross-point defects, giving rise to growth, shrinkage, appearance and disappearance faults. Once the model has been established, it is not difficult to identify test vectors for which fault-free and faulty circuits will give different responses.

ROMs of all kinds, whether mask-programmed, field-programmable, or erasable and reprogrammable, are simple to test in that there is no alternative to an exhaustive test. This requires a substantial test application time which is unavoidable, but the test data, in the form of the set of fault-free responses together with the responses collected from the UUT, can usefully be compressed using either check sums or CRC signatures.

RAMs have a complex structure and are subject to a wide variety of defects, including particularly those that give rise to pattern-sensitive and soft errors. Test procedures for RAMs have been devised at various levels of complexity; the most comprehensive such procedures, typified by GALPAT, takes a time proportional to the square of the number of cells, and is therefore not a practicable proposition for modern large memories (64K and above).

Microprocessors, as the most complex devices at present on the market, are the most difficult to deal with. Some form of functional test strategy has to be employed, and several such strategies are briefly described.

EXERCISES
Chapter 6

E6.1 The circuit shown consists of three 2–1 multiplexers and an inverter, and represents part of a pcb populated with TTL SSI and MSI chips. As a preliminary to writing a test program for this circuit, a test engineer compiles a fault-list consisting of each primary input and each element output s-a-0 and s-a-1 (a total of 14 faults). Give examples of physical defects whose effects might not be covered by this fault list.

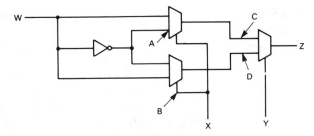

It is anticipated that, during manufacture of this pcb, the following defects might occur:

a a dry joint between the ic pin at point A and the pcb track;

b a short circuit between the track at point B and ground;

c a solder splash between tracks at points C and D.

Write tests to reveal each of these defects, explaining what fault-model you use and how you derive the tests. Establish the full single-stuck fault-cover for each test that you write. [Assume that for each multiplexer a 0 on the select line connects the output to the upper input, e.g. if $Y = 0$, then $Z = C$.]

E6.2 A certain TTL chip contains four half-adders, each of which has two inputs, X and Y, and two outputs, S (sum) and C (carry). This chip is used with ancillary gating to implement the two-bit full-adder shown.

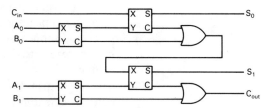

Assuming that the only access to the circuit is through the primary inputs (C_{in}, A_0, B_0, A_1, B_1) and the primary outputs (S_0, S_1, C_{out}), develop a test set for this circuit, to cover all single-stuck faults. Explain carefully the derivation of each of your tests.

E6.3 The logic block shown is a full-adder. What input conditions are necessary

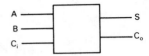

a to establish fixed values at the outputs;
b to establish fault-sensitive values at the outputs;
c to transmit fault-effects through the block?

E6.4 Explain what is meant by a cross-point defect in a PLA. Show what possible fault-effects can be produced by cross-point defects at different places in the device.

E6.5 A particular PLA of the 'blown diode' type is shown diagrammatically below.

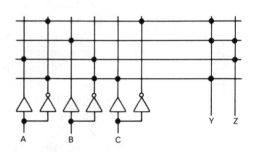

a What product terms are programmed?
b What are the output functions?
c Identify a crosspoint defect that will produce
   i   a growth fault;
   ii a shrinkage fault;
   iii an appearance fault;
   iv a disappearance fault.
d Derive tests to cover each of the faults identified in **c**.

E6.6 The mechanism for selecting a location in a ROM consists essentially of a decoder. What faults in this decoder need to be considered when preparing a test program? How would these be covered?

E6.7 A test programmer devises a test for a RAM on the basis of covering all single-stuck faults on the I/0 pins. Comment on the likely effectiveness of such a test program.

E6.8 The 74189 is a 16 word × 4 bit RAM. A possible test program takes the following form:

1 Write 0 into all locations.
2 Read 0 from all locations.
3 Write 1 into all locations.
4 Read 1 from all locations.

If the address decoder has a defect that results in the LSB being s-a-0, show that this test would still be passed.

Show further that the MARCH algorithm (see section 6.4.3) would detect this fault.

E6.9 One test of the program counter in the 6800 microprocessor is to check that it can increment throughout its range. Suggest a procedure to achieve this. (Consider the action of the NOP instruction.)

7

ENHANCING TESTABILITY

7.1 ASSESSING TESTABILITY

7.1.1 The meaning of testability

In the course of developing test programs for various circuits, it will have become apparent that two circuits of similar complexities can present test problems of very different orders of difficulty. This gives rise to the concept of the testability of any particular circuit; we require a quantitative measure that indicates the ease or difficulty of deriving a test program for that circuit. The concept is straightforward and easy to appreciate; the production of a definition of testability and a set of rules for calculating its value is a much less straightforward matter.

Once a test program has been produced, it is common to count the number of faults covered by the program and to express this as a percentage of the total number of faults. This number is then often used as a figure of merit, and is often described as a measure of testability. There are, however, several reasons why this cannot be regarded as satisfactory, either as a definition of testability or as a measure of the quality that we wish to assess:

a A simple binary score attached to each fault (covered or not covered) takes no account either of the difficulty of deriving the test or of the time and effort required to apply it. In other words, while a fault for which no test can be found is clearly untestable, it is equally clear that there are important differences between two implementations of a circuit, one of which requires 100 patterns to cover 95% of all single-stuck faults while the other requires 1000 patterns to achieve the same thing. The larger program will have cost more to produce, whether manual or computer based TPG methods were used; it will also cost more to apply, requiring more time per unit on the ATE. It is clear that an analogue measure will be needed if the desired distinctions are to be made.

b A more fundamental objection to the use of fault-cover as a measure of quality is that it depends on the fault-list used. Faults that are not included on the list will not contribute to the measure however important such faults may actually be in the particular circuit. The faults may, of course, be covered fortuitously, but if they are not, and if it is difficult or impossible to derive appropriate tests, then this should be reflected in a testability measure.

c One of the most important uses for which a testability measure is required is to assess the quality of a particular design or implementation from the point of view of testing. The objective here would be to detect difficulties at a stage at which they could be alleviated by modifications to the design. In view of this requirement, it is clearly undesirable that a complete test program should have to be derived before the testability measure can be calculated.

d Even if an analogue version of fault-cover could be devised to satisfy the objections noted under **a** above, there would still remain serious doubts about its usefulness as a figure of merit. The doubts may be illustrated by consideration of Fig. 7.1, which represents a general logic circuit with one particular internal node identified. Let us suppose that this node is fully testable (that is, we have tests for the node s-a-0 and s-a-1), but that there are faults in other parts of the circuit that are not covered, so that the overall fault coverage is less than 100%. If now the circuit is modified, as shown in Fig. 7.2, by the insertion of two (or any even number) of inverters, we will have a circuit with the same logical function, but with extra propagation delays, higher cost, and lower reliability (because extra components means more things that can go wrong). By any criterion, therefore, the circuit of Fig. 7.2 is inferior to that of Fig. 7.1; in terms of fault-coverage, however, Fig. 7.2 rates higher than Fig. 7.1 with no change in the test program, because faults on the introduced nodes are all covered, being indistinguishable from those on the original node. This example may seem artificial, but the author knows of a case in a major manufacturing company in which exactly this modification was made to a circuit in order to make it conform to a fault-coverage specification.

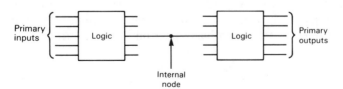

Fig. 7.1   The testing problem related to an internal node.

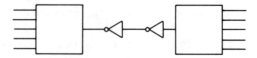

Fig. 7.2   Making the circuit poorer but the fault-cover higher.

Consideration of all these points leads us to the conclusion that fault-coverage is not suitable as a measure of testability. The unsuspected complications attached to what at first sight appears to be a simple and straightforward measure also serve as a warning that the derivation of an acceptable measure (that is, one that does not offend against intuitive notions of testability rankings) is likely to give rise to controversy.

7.1.2   Controllability and observability

If we consider any particular node in a circuit, such as the one illustrated in Fig. 7.1, we know that there are two distinct stages in the derivation of a test for the node stuck at one particular value. The first stage is to set up the appropriate fault-free value at the node by applying a set of values at the primary inputs; the second stage is to transmit the actual value at the node to a primary output. The ease with which each of these stages can be accomplished depends essentially on the components and the topology of the circuit; this leads to the concept of two significant properties that can be iden-tified at each node of the circuit:

a	controllability	the ability to control the fault-free logic value at the node from the primary inputs;
b	observability	the ability to cause a change from the fault-free value at a primary output to result from a change from the fault-free value at the node.

The key to testability lies in the combination of these two properties; the inability to achieve either stage of the process for any particular fault will mean that the fault is untestable. Most testability measures that have been proposed start by defining measures of controllability and observability; these in turn are used to calculate testability. The measures are defined in such a way as to satisfy the intuitive understanding of structures that are easier or harder to test; they have no absolute significance, but serve to guide circuit designers who can make modifications to sections of their circuits that are likely to cause testing problems later on.

7.2 STRUCTURES WITH POOR TESTABILITY

7.2.1 Circuits with redundancy

It was shown in Chapters 3 and 5 that the existence of logical redundancy in combinational circuits, and of redundant states in sequential circuits, could result in some faults being undetectable. It was also pointed out that some redundancy is almost inevitable in practice because of the use of modular techniques in combinational logic design, and because the number of states in a sequential circuit has to be a power of two. Where redundancy results in undetectable faults, this can be viewed as an example of fault-tolerance, in that the existence of that particular fault will have no observable effect on the performance of the circuit. This in turn might suggest that such faults are of no operational significance and can therefore be safely ignored; however, as with most facets of the testing problem, the matter is not quite as straight-forward as it at first appears.

Before deciding on the importance or otherwise of an undetectable fault we really have to consider again the fault model that we are using. If, as is usually the case, the analysis is based on the single-stuck fault model, we are assuming not only that the node in question has a constant logical value, but also that there is no other effect on the operation of the rest of the circuit. For some defects, this is a reasonable assumption; an open circuit at the input to a TTL chip, for example, will almost certainly do nothing except make the node appear s-a-1. However, consider a gate with a short-circuit across one of its transistors such that the output is s-a-1. In this case there may be a number of consequential effects apart from the stuck logic level; these all amount to performance degradation rather than outright failure:

a The current drawn by the gate from the power supply could be substantially higher than for the fault-free circuit, resulting in higher power consumption and (more importantly) greater heat generation. An increase in temperature in the system will certainly not improve its performance, and is likely, sooner or later, to lead to breakdown.

b The output voltage appearing at a s-a-1 node may well be significantly different from a normal logic 1 level (e.g. 5 V instead of 3.5 V). This could affect the performance of the stage following the faulty one in terms of current requirements and propagation delay. Timing changes (which are not represented in logic equations) can be particularly insidious in their effects, since they can introduce glitches, but these glitches will occur only with particular data patterns, and will in any case not necessarily affect the output. These are intermittent faults, which are notoriously difficult to detect, diagnose, and repair.

c A faulty gate can present a larger than usual load to the output of a preceding gate. This affects the fan-out, and can again cause intermittent failure if it occurs at a point in the circuit that is already operating close to its permitted limit.

These effects may seem rather nebulous, and it is certainly true that the presence of undetectable faults might not cause any difficulties. It should, however, be noted that if side effects such as those suggested above do occur they are most likely to be seen at a late stage in the manufacturing process, most probably in the field. The effects are all also likely to be intermittent and very difficult to find (for example, when the system is opened up for servicing, temperature effects tend to disappear because of the increased ventilation). The conclusion to which we are driven is that all faults in a circuit are undesirable, even if they do not apparently affect the logical operation of the circuit.

There is an additional hidden cost attached to the presence of an undetectable fault. The process of test pattern generation, whether performed manually or automatically, is essentially a searching process based on trial and error. An undetectable fault will lead to an exhaustive search before undetectability is finally established; this is a time-consuming, and therefore expensive, process.

Undetectable faults can usually be made detectable by increasing the observability or controllability or both, using techniques such as those described in section 7.3. (See also section 2.4.)

7.2.2 Asynchronous design

The difficulties attached to the use of asynchronous design have already been discussed (see Chapter 5). It is generally considered good design practice to stick to synchronous systems wherever possible; this has benefits in terms of reliability of operation as well as making the function of the system much more understandable. From the testing point of view, asynchronous design is particularly bad when diagnostic test programs are required. From all points of view, therefore, asynchronous circuits should be avoided.

7.2.3 Unstructured synchronous design

Avoiding the use of asynchronous logic is not enough. As the example in Chapter 5 showed, even a simple synchronous circuit can cause enormous difficulties when it comes to test pattern generation. This is particularly so in the context of automatic (that is, computer-based) TPG; algorithms that are capable of dealing with stored-state devices tend to be slow in execution and uncertain in effectiveness.

The difficulties of dealing with sequential circuits are so serious that it has become widely accepted, particularly among major manufacturers of mainframe computers, that it is necessary to adopt a completely new approach to circuit design, restricting the designer to using structures that can easily be tested. The concept of design for testability is so important that it will be dealt with separately in Chapter 8.

7.3 IMPROVING ACCESS

7.3.1 Extra input/output pins

The most straightforward approach to counteracting poor controllability or observability at an internal node is to provide direct access between the node and an external connection point (primary input or output). This can be achieved in various ways:

a If there are unused edge-connectors on a pcb, or unused pins on a chip, these can be used simply to bring out internal nodes for observation. To provide direct control to an internal node generally requires two pins: one for the tester input and another to disable the normal control. In the circuit of Fig. 7.3(a) for example, external inputs A and B have been added to the logic that normally drives node Z. When $AB = 10$, the function is unchanged from the original; but if $A = 0$, the normal inputs are disabled and $Z = B$. If the node of interest is driven by logic that is inaccessible, and cannot therefore be modified, a similar effect can be obtained by inserting a pair of NAND gates into a connecting link, as shown in Fig. 7.3(b). This technique, whereby a circuit can be electronically split for testing purposes into separate and independent parts, is known as **degating**.

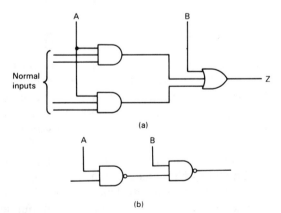

(a)

(b)

Fig. 7.3 Electronic partitioning. (a) Degating using extra inputs on existing gates; (b) degating by introducing additional gates.

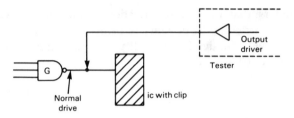

Fig. 7.4   The problem of backdriving.

b On a pcb, additional access can be provided by fitting special contact points solely for connection to the test equipment. These can take the form of:

i stake pins (terminal posts), to which individual probes can be connected;

ii dual-in-line sockets, providing clusters of access points;

iii additional edge-connectors (perhaps along one of the other edges of the board).

c Also on a pcb, it may be possible to attach individual probes or ic clips to particular nodes or components. These can be used without restriction for observation, and to a limited extent for control. The limitation here is because, as shown in Fig. 7.4, we are really connecting two outputs together; the output drive from the tester and the normal drive from the circuit (gate G in Fig. 7.4). If the circuit is trying to drive the output of G to one state while the tester is trying to drive the same point to the other state (a process known as **backdriving**), then this could result in irreparable damage. The form of the limitation depends on the technology in use; with TTL, for example, gate outputs can be safely pulled low but not high.

7.3.2   Multiplexers and shift registers

Even with pcbs, it will often be inconvenient or uneconomical to provide all the extra access that may be desirable from the testing point of view; with chips it is usually impossible. One method of overcoming limitations on the number of available pins is to use multiplexers to allow existing primary inputs and outputs to be assigned an alternative function for testing purposes.

The principles are illustrated in Fig. 7.5; one control pin (M_1) can select any number of internal nodes to be switched from their normal drive to direct drive from primary inputs; another control pin (M_2) can select any number of primary outputs to be switched from their normal circuit outputs to observe internal nodes. With this arrangement we have three useful modes of operation:

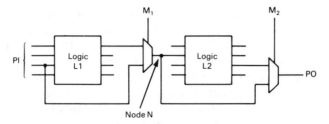

Fig. 7.5 The use of MUXs to allow shared used of I/O pins.

$M_1 M_2$ = 00 : Operational mode
$M_1 M_2$ = 01 : Test L_1 – observe N at PO.
$M_1 M_2$ = 10 : Test L2 – control N from PI.

Another way of economizing on primary output pins is to use one pin to observe a number of internal node values in turn. This can be done by incorporating a shift register with parallel load facility to capture the required data, shifting it out through a single pin for observation. In a similar way, data can be shifted serially into a shift register through a single primary input for subsequent use as a (parallel) control vector. In each case, some control pins will be needed in addition to the single data entry/exit pin.

7.4 DISABLING PATHWAYS

7.4.1 Feedback paths

The use of asynchronous design, characterized by global feedback connections around blocks of combinational logic, is recognized by most authorities to be very undesirable. Many CAD systems, in fact, have design rules that prevent designers from making use of this kind of construction. If, however, global feedback is used, the testing problem has to be recognized and solved.

It is when attempting to perform diagnosis that the main limitation of a circuit with feedback becomes apparant. The usual basis of diagnosis, as discussed in Chapter 4 (see section 4.2), is to identify a component whose inputs are correct but whose output is wrong; this is assessed on the basis of a test program, using a compressed data signature.

If a fault develops in a component that is contained within a loop, then, as discussed in section 4.5, wrong signatures will be propagated all round the loop so that diagnostic information is not available. Considerable research effort has been devoted to the loop breaking problem, but no satisfactory solution has so far been found. The only way of avoiding this problem is to make provision for the breaking of feedback loops for testing purposes.

Feedback loops on pcbs can be broken using either mechanical or electronic methods. Mechanical methods involve producing a physical break in
the signal path; essentially this is achieved by incorporating the break into the
basic structure, with a short circuit link (a jumper) which can be removed for
testing. This arrangement not only breaks the loop but also provides a point
for additional control. It is particularly convenient to route a feedback path
through the edge-connector if there are two edge-connector positions
available. These two positions are short-circuited at the socket in the system;
the pathway is automatically broken when the pcb is removed and put in the
tester.

Electronic methods of breaking feedback paths, which are the only
methods possible when dealing with single chips, are all variations of the
degating technique described in section 7.3.1, and depend on incorporating a
control signal so that the pathways can be disabled by an external input. Any
number of pathways can be disabled by a single control signal, but if direct
control of the logic level is required then further inputs, and perhaps additional gating, will need to be employed using techniques such as those illustrated in Fig. 7.3.

7.4.2 Asynchronous inputs

The practice of applying internally generated logic signals to the asynchronous inputs of flip-flops, as shown in Fig. 7.6(a), is not recommended; it
has the effect of introducing asynchronous behaviour with its inherent risk of
hazards. If it is considered essential to use a circuit with this structure, then
the testing problems can be reduced by providing a facility, such as shown in
Fig. 7.6(b), to allow the asynchronous input to be disabled by the tester.

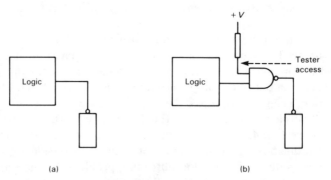

(a)   (b)

Fig. 7.6 Asynchronous inputs driven by system logic. (a) The simple
connection, which presents testing difficulties; (b) a testable alternative.

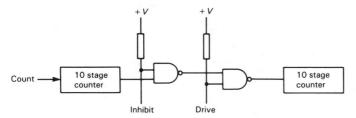

Fig. 7.7   Use of degating to simplify the testing of long counter chains.

7.4.3   Counters and oscillators

In order to test a counter it is certainly necessary to make it count through its range. It may also be necessary, when setting up tests for other parts of the circuit, to drive the counter to particular values. The problem here is that a long counter will take a long time to step through its states; the problem can be alleviated by dividing the counter into sections using degating. The normal form of this is shown in Fig. 7.7; the degating is essentially the same as that shown in Fig. 7.3(b), but the addition of the bias resistors means that, in normal use, no connection is needed to the test inputs. Operationally, the circuit of Fig. 7.7 acts as a twenty-stage counter; but whereas a single twenty-stage counter requires over a million pulses to step it through its range, the degated version can be tested as if it consists of two independant ten-stage counters each of which requires only about a thousand pulses to step through its range.

The use of free-running oscillators (usually used as on-board clocks) poses a different problem for testing. This problem is essentially one of synchronization between the tester and the circuit; without this synchronization it is difficult, if not impossible, for the tester to maintain control over the execution of the test program. The solution to this problem is to make provision for the on-board clock to be disabled, and for the required clock signals to be supplied to the rest of the circuit by the tester. This can be achieved either by the use of a removable link or by the use of a degating circuit.

7.5   PARTITIONING

Many of the problems associated with testing arise from the exponential relationship between the number of inputs to a circuit and the number of possible test vectors. This directly affects the process of test pattern generation; the cost of the process, whether performed manually or automatically, is usually considered to be proportional to at least the square (some say the cube) of the number of inputs. Thus, if a circuit with N inputs can be partitioned for

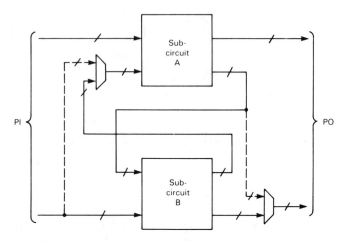

Fig. 7.8 Partitioning a circuit by the use of MUXs.

testing purposes into M sub-circuits each with N/M inputs then the total TPG cost is likely to be reduced in the proportion

$$M \times \left(\frac{N}{M}\right)^2 \div N^2 = \frac{1}{M}$$

Partitioning can be achieved by using multiplexers to break the interconnections between sub-circuits and to provide the necessary control and observation. The general principle is illustrated in Fig. 7.8, which shows a circuit with two sub circuits, A and B, each driven by some primary inputs; sub-circuit A is also driven by some inputs derived from sub-circuit B, and sub-circuit B is driven by some inputs derived from sub-circuit A. Similarly, each sub-circuit generates some primary outputs in addition to the outputs feeding the other sub-circuit. By incorporating the two multiplexers shown, and adding the pathways shown dotted, sub-circuit A can be tested in isolation from sub-circuit B, 'borrowing' as many of the primary inputs and outputs of sub-circuit B as are needed to complete its observation and control. Similar modifications can be made to allow sub-circuit B to be tested in isolation from sub-circuit A.

7.6 INITIALIZATION

The importance of initialization in the testing process can scarcely be over-emphasized. The most straightforward way of providing for every node in the circuit to be set to a known state irrespective of its starting state is to provide a single master reset input which drives every flip-flop (and other sequential elements) through its asynchronous inputs. There is no need for all flip-flops to be driven to the same state; the only requirement is that they should be driven to a known state.

If there is no master reset, a flip-flop will have to be initialized either through its data inputs or by attaching a clip to one of its asynchronous inputs. The latter option, however, can be difficult or impossible to apply with certain circuit configurations. Direct connection of the asynchronous inputs to a power rail, as shown in Fig. 7.9(a), is usually regarded as poor practice for operational reasons; it certainly prevents the tester from gaining access to them. Scarcely better is the inclusion of a single resistor driving both inputs, as shown in Fig. 7.9(b). Here the problem is that to pull both asynchronous inputs low simultaneously by attaching a probe to one of them will cause an indeterminate transition of the flip-flop. The initialize-on-power-up circuit shown in Fig. 7.9(c) has a different problem: a probe could be attached to the asynchronous input, but the time required to discharge and charge the capacitor is likely to be several orders of magnitude larger than the normal interval between tests.

The solution to these difficulties is straightforward: if a master reset is not provided, then at least there must be provision for the tester to be able to override the normal circuit function. Figs. 7.9(d) and (e) show acceptable alternatives to the unsatisfactory arrangements of Figs. 7.9(b) and (c).

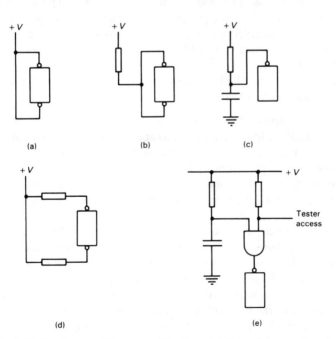

Fig. 7.9 Initialization problems on a flip-flop. (a) Asynchronous inputs not controllable by tester; (b) initialization state not definable by tester; (c) preset (or clear) on power-up difficult for tester to override; (d) tester-controllable alternative to (a) and (b); (e) tester-usable alternative to (c).

7.7 PROBLEMS WITH TESTING PCBs

7.7.1 Physical access

The suggestions advanced in the earlier sections of this chapter have concentrated on ways in which circuit design should be constrained (or modified) so as to make tester control possible. Consideration has been restricted so far to the electrical properties of the circuit; but there is also a need to take account of the physical layout.

The circuit designer is continually under pressure to provide increased capability from his system, but, at the same time, market forces demand competitive pricing. Since a significant fraction of the total system cost is accounted for by mechanical parts (such as pcbs, plugs and sockets, housing, etc.), a system fitted on to three pcbs is likely to be considerably cheaper than the same system on four pcbs; the incentive to maximize the packing density of components on a pcb is clear.

The conflict with the requirements of the test engineer is now apparent: the methods of achieving tester control involve, in many cases, attaching probes or clips to the circuit; even if the main test program accesses the circuit only through the edge-connector, it will almost certainly be necessary to probe to achieve fault diagnosis. The ease with which the tester can gain access to the circuit will be affected by the physical arrangement of components on the board; from the manufacturer's point of view it is worth remembering again that anything that makes testing difficult may prove, in the long run, to be a false economy.

Among the features of physical layout and construction that cause difficulty we may note the following:

a Access by clip and probe can be made impossible if components are too close together. In particular, the temptation to slip discrete components (such as pull-up resistors and decoupling capacitors) into small gaps between chips should be resisted.

b It is not usually easy, and is certainly not usually convenient, to probe both sides of a pcb. Components mounted on the 'wrong' side of the board make for complications.

c Multi-layer boards make for difficulties particularly in using current probes. Burying of high fan-out nodes should be avoided if possible. For the same reason, track junctions should not be hidden under ics.

d The advent of surface-mounted and pin grid array devices promises to produce new access problems, with some of the device pins being hidden under the body of the device. Where these pins represent critical nodes in the circuit it may be necessary to bring them out to probe points for tester access.

e Diagnosis and repair can be made much easier if ics are fitted in sockets. Unfortunately, there is a substantial price to be paid for this convenience: sockets are expensive (in the case of common SSI chips the socket costs more than the chip); they appreciably increase the physical size and weight of the circuit; and they introduce the likelihood of problems with poor contacts (particularly undesirable because these are most likely not to appear until the equipment is in the field). Despite all these objections, the designer should at least consider the possibility of socketing major components such as microprocessors.

7.7.2 Electrical interface

The attaching of a probe to any node in the circuit, whether at an edge-connector or at a component pin, will impose an electrical load. It is, therefore, unwise for the circuit designer to allow any node to be loaded internally up to the full fan-out limit, as there would then be a risk of malfunction under test conditions.

Open-collector gates can be a source of trouble, because of the need to connect pull-up resistors for correct operation. The difficulty is that the pull-up resistor may not be on the same board as the gate. Some testers have a built-in facility for attaching pull-up (or pull-down) resistors to its inputs; if this is not available, the necessary resistors will have to be supplied as part of the tester interface.

7.7.3 Difficult components

From the testing point of view, it is particularly difficult to deal with circuit elements whose operation is essentially dependent upon, or characterized by, an absolute time scale. This includes most analogue components, but it also includes at least two digital components: monostables and dynamic RAMs.

Monostables are characterized essentially by a transitory logic change, which is difficult for the tester to catch; also an important performance criterion is the time interval between transitions, which is difficult for the tester to measure. The actual width of the monostable pulse is also a restriction; it will often be very long relative to the normal clock period so that the test program can be made unacceptably long if it has to include operations of the monostable. At the other extreme, the monostable period can be much too short for convenience. The solution to all these problems lies in direct tester access: the monostable output needs to be observable. To overcome problems with inconvenient pulse width it is possible to modify the timing components: an additional capacitor will stretch the pulse, while an additional resistor will shorten it. Alternatively we can use degating at the output (as with an on-board clock) so that the monostable can be removed and its function assumed by the tester. Ultimately the best advice with monostables is not to use them!

Dynamic RAMs (and also other dynamic logic) have the inconvenient requirement for refresh pulses at intervals of not more than a few milliseconds. If these pulses have to be supplied by the tester, it puts severe constraints on the test program. The most straightforward solution to this problem is to ensure that all refresh circuitry that is needed is provided on the board with the components that need it.

7.8 COSTS OF TESTABILITY IMPROVEMENT

It is important to observe that all the methods that have been suggested so far for improving the testability of a circuit have incurred costs; it will become apparent that this is invariably the case whenever design is constrained by testability requirements. The economic arguments discussed in Chapter 1 have to be invoked to assess the extent to which it is worth while applying those techniques in any particular case.

There are at least four ways in which testability enhancement incurs penalties; both production costs and operational performance can be affected:

a Whatever techniques are applied, there is always a need for some additional input/output pins, either for direct access or to allow control of multi-function operations. The hardware cost of providing these pins is not limited to the cost of the pins themselves: additional connections to the ATE imply additional interface costs and also additional running costs because of the extra set-up time.

b Where extra circuitry, such as buffer gates or multiplexers, is included solely for test purposes there are obvious demands on board or chip area (which is often under heavy pressure from the circuit designer). Moreover, even when there is no call for extra circuit elements, board area will still be consumed by additional wiring.

c The inclusion of extra gates in the signal paths will imply additonal propagation delays, and hence a degradation of the operational performance.

d If there are more components in a circuit, there are more things to go wrong. Hence we must expect a reduction in reliability for the circuit as a whole.

It is against this background that we might reconsider our whole approach to circuit design, and, rather than improving testability after the circuit has been designed, attempt to build testability into the circuit from the start.

SUMMARY
Chapter 7
Because of the importance of testing in the manufacturing process, and of its contribution to the cost of the

final product, there is a need to be able to assess the quality of a circuit by formulating a measure of testability. However this is defined, it depends on the controllability and observability of the internal nodes; by improving these aspects of the circuit design, the overall testability will be enhanced.

Redundancy in a design, which might be incorporated deliberately, or might be the result of a non-minimal implementation, will often result in undetectable faults. These can produce unwelcome side-effects, which can be avoided either by removing the redundancy or by making the faults testable.

It is particularly important for the tester to be able to control elements such as flip-flops and counters for initialization purposes; for this to be possible, the asynchronous inputs of these components must be available to the tester, and they must also not be hard-wired to their inactive states. Access to these and other internal circuit nodes can be improved by the use of extra pins, edge-connector fingers or stake pins, or by incorporating multiplexers or shift registers to permit the sharing of available pins between circuit and test functions.

If the process of TPG is to be made reasonably straightforward, complex circuits need to be modified in various respects to prepare them for testing. This reconfiguration can be achieved, for example, by breaking feedback paths, and partitioning large circuits into blocks of more manageable size, using techniques such as degating.

There are always cost penalties attached to testability enhancement: as well as having purely financial implications, modifications often also bring with them marginal degradation of performance and of reliability. These costs cannot be avoided; if the circuit costs are minimized by ignoring testability requirements, the capital costs are simply transferred to testing costs, and are probably amplified in the process.

EXERCISES
Chapter 7

E7.1 Explain the concepts of observability and controllability and their significance in relation to the problems of testing electronic circuits.

E7.2 If a circuit has an undetectable fault, then by definition there is no input or sequence of inputs that will produce any output other than the fault-free one. In that case, should the test engineer be concerned that the fault remains undetected by his test program?

E7.3 Explain what is meant by degating, illustrating your answer with circuit diagrams, and describe the circumstances in which it is used. Discuss the benefits and costs of incorporating degating within digital circuits.

E7.4 To gain access to significant numbers of internal nodes for testing purposes can require more I/O pins than are available on the chip or pcb. Discuss the possible ways in which this problem can be addressed.

E7.5 A certain combinational circuit has 20 primary inputs and 8 primary outputs, and is mounted on a pcb with a maximum of 32 edge-connector fingers, of which 2 are required for the power supply. A study of the circuit reveals that it can be regarded as two sub-circuits as shown below.

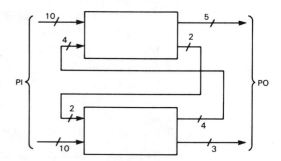

Explain how this circuit can be modified to make testing easier. Compare the times needed to conduct an exhaustive test for the original and the modified circuits.

If the total circuit originally contained the equivalent of 1000 gates, estimate the hardware overhead of this circuit modification.

8

DESIGNING FOR TESTABILITY

8.1 LIMITATIONS OF TESTABILITY IMPROVEMENT SCHEMES

The techniques outlined in Chapter 7 for improving the testability of a circuit are helpful as far as they go, and they are generally recognized in the Industry as constituting good design practice, but in themselves they do not provide the complete answer to testability problems. The difficulty is that most of the guidelines really amount to suggestions for taking a design that is basically unsatisfactory from the test engineer's point of view, and modifying it so as to reduce the worst effects. In doing this, we incur various additional hardware costs, and also come up against physical restrictions such as the limitation on the number of input/output pins that can be accommodated on the board or chip. **Design for Testability** (DFT) is a fundamentally different approach, in which the circuit is designed from the start in such a way that testing problems will not arise.

The basic problems of testability can be illustrated by reference to the general model of a sequential circuit shown in Fig. 8.1. This model consists of three parts: input logic, output logic, and flip-flops.

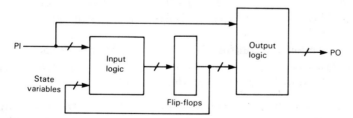

Fig. 8.1   A general model of a synchronous sequential circuit.

a The input logic is a block of combinational logic that provides the control signals defining the next states of the flip-flops (it is often known as 'next-state logic'). These signals are functions of the present state and the primary inputs.

b The flip-flops define the state of the system.

c Outputs are derived by the output logic as combinational functions of the primary inputs and the state variables.

From this model, it is not difficult to identify the features in the structure that are likely to cause problems in testing:

a To test the combinational logic, we need to apply specific test vectors. But each block has as some of its inputs the state variables, over which we have no direct control.

b Having applied the test vectors to the input logic, the outputs are not directly observable.

c The state variables themselves (that is, the outputs of the flip-flops) are neither directly observable nor directly controllable.

To overcome these problems of controllability and observability we could contemplate using the techiques outlined in Chapter 7, but for a circuit of any realistic size the chances of being able to accommodate all the additional input/output pins are small indeed, even if the cost were acceptable. The objective of structured design for testability is to obtain complete controllability and observability without requiring more than a minimal increase of input/output pins. To achieve this objective requires adherence to a design discipline; it is not possible to tinker with an already existing circuit, because it is not only necessary to adopt specific architectural features, but also in most cases to use specially tailored components.

8.2 THE SISO PRINCIPLE

All the formal methods of design for testability that have so far been proposed depend on the same two basic principles:

a The structure of the system is such that the stored-state devices and the combinational logic can be isolated from each other for testing purposes.

b Observation and control of the state variables are both achieved using serial access so as to minimize the number of input and output pins that have to be dedicated to the testing function.

It is because of the serial access feature that these are known as **scan-in, scan-out** (SISO) methods and the serial data path is known as a **scan-path**.

To see how a SISO system is related to the general synchronous sequential

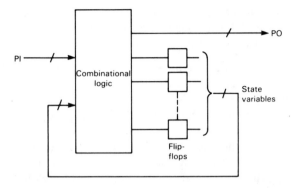

Fig. 8.2   Compacted form of Fig. 8.1, emphasizing the two parts of the circuit.

system shown in Fig. 8.1, we should first note that both the input logic and the output logic have the same set of inputs, made up of the primary inputs and the state variables. We can, therefore, conveniently lump all the combinational logic into a single block as shown in Fig. 8.2. We assume, for simplicity, that each stored-state device is a D-type flip-flop, and that they are all operated by a single clock.

The modification necessary to made this into a SISO system is shown in Fig. 8.3, and consists essentially of inserting a multiplexer in front of the data input of each flip-flop, all the multiplexers being controlled by a single mode control M. If $M = 0$ the upper inputs are selected, while if $M = 1$ the lower inputs are selected. The output of each flip-flop, as well as being connected to the combinational logic, is also connected to the next flip-flop through its multiplexer. A complete chain of flip-flops is thus formed, and the two ends

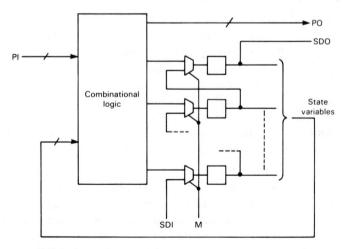

Fig. 8.3   The SISO principle.

of the chain are brought out as input/output pins; these are generally known as the **scan data in** (SDI) and **scan data out** (SDO) pins.

The system has two modes of operation, selected by the mode control M.

a When $M = 0$, the system is in operational mode, with the combinational logic driving the flip-flops.

b When $M = 1$, the system is in scan mode, with the combinational logic disconnected from the flip-flop inputs. The flip-flops are reconfigured into a single shift-register with input (SDI) and output (SDO) directly available. Thus the state of the circuit is directly controllable, by shifting in any required set of values through SDI; and directly observable by shifting the contents of the flip-flops out through SDO.

With this circuit arrangement, test pattern generation is greatly simplified compared with an unstructured design. The set of flip-flops can be treated as a shift register with direct access to input and output. The combinational logic can be considered in isolation, with control of all inputs and observation of all outputs.

The test procedure with the SISO structure can be summarized as follows:

1 Test the flip-flops as a shift register, by setting $M = 1$, and using SDI and SDO for access. One possible test is to clock through the sequence 00110011------. This requires each flip-flop to make each of the four possible transitions. Notice that, whatever test is devised for a shift register, it will serve for all circuits, since flip-flops are always configured as a shift register for test purposes, the only variation being in the length.

2 Test the combinational logic. Tests will have been derived assuming controllability of all inputs (state variables as well as primary inputs) and observability of all outputs (flip-flop inputs as well as primary outputs). For each test there is a three stage process.

a Set $M = 1$ and shift in the required set of state variables.

b Set $M = 0$. Apply the required primary inputs. Observe the resulting primary outputs. Clock the flip-flops once, so that the remaining outputs from the combinational logic are latched into the flip-flops.

c Set $M = 1$. Shift the flip-flop contents out and observe them at SDO. In practice, steps (a) and (c) can be combined, in that the state variables required for test $N + 1$ can be shifted in at the same time that the results from test N are shifted out.

In Fig. 8.3, three extra input/output pins are shown. This is the maximum that would be required; in fact, if pin-count is a serious limitation (as it might well be in a chip) this number can be further reduced. A study of the test procedure outlined above will reveal that the times at which the scan pins (SDI and SDO) are required never overlap with the times at which the primary inputs and outputs are required. It is therefore possible for SDI to share with

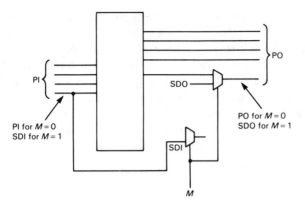

Fig. 8.4   Using MUXs to reduce the number of I/O pins required in a SISO system.

a primary input with no cost apart from the wiring, and for SDO to share with a primary output at the cost of a multiplexer. These modifications are shown in Fig. 8.4.

8.3   LEVEL SENSITIVE SCAN DESIGN

8.3.1   The shift register latch

The most widely known implementation of SISO is **level sensitive scan design** (LSSD), which is widely used by IBM, and has been heavily promoted by them. Many other manufacturers have used LSSD or variations of it.

Most implementations of SISO depend on the use of a specially designed flip-flop that effectively combines into a single unit a storage element and a selection mechanism (equivalent to a multiplexer); the LSSD version of this unit is known as the **shift register latch** (SRL). The designers of the SRL were conscious not only of the need to provide a scan path facility, but also of the difficulties posed by circuits that are sensitive to timing constraints. The operation of such circuits can be critically dependent on the rise or fall times of clock pulse edges, and where the circuit has multiple inputs, the response to nominally simultaneous input changes can also be affected by the exact order in which input values actually change, which in turn will be dependent on propagation delays in gates and transmission paths. Since all these ac characteristics of a circuit are impossible to control or predict accurately such dependence is very undesirable.

The SRL overcomes the problems of sensitivity to ac parameters by using a master/slave construction, where each section of the flip-flop is a level sensitive latch. Data is latched into each section by a clock signal, the two clocks being independent and non-overlapping. A gate level equivalent of the SRL is shown in Fig. 8.5(a); the main master/slave flip-flop consists of gates 3,4,7,8 (master) and 9,10,11,12 (slave). The data input is D, and C and B are the two phase clock inputs (which are common to all the SRL's). The scan-path facility consists of gates 2,5,6; this provides an alternative front end to the master flip-flop, with its own data input (I) and its own clock (A). The SRL can be shown diagrammatically as in Fig. 8.5(b); the master and slave sections are indicated by L1 and L2. Notice that we have the option of taking an output from L1 instead of (or as well as) L2.

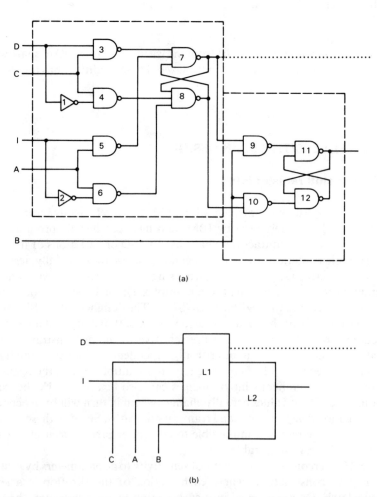

(a)

(b)

Fig. 8.5 The shift register latch. (a) Circuit diagram; (b) block representation.

8.3.2 Double latch LSSD

The most straightforward way of implementing LSSD using SRLs is illustrated in Fig. 8.6. The three clock inputs, C, A and B, are common to all SRLs. For normal system use, data is clocked into the SRLs using clocks C and B, and the L2 outputs form the state variables of the system. The master/slave nature of the SRLs ensures that the system cannot suffer from race problems. The scan path is formed by chaining the SRLs with the L2 output of each SRL connected to the I input of the next SRL in the chain; the scan path is highlighted by the solid line in Fig. 8.6. Data is shifted through the scan path by using clocks A and B.

The method of using the LSSD system is essentially the same as that described in section 8.2, except that instead of using a mode control input to select between system mode and scan mode, the choice is made by activating either clock C or clock A as the first phase of the two-phase clock. The test procedure for each combinational logic test vector is, therefore:

1 Shift in the required state variables through SDI, using clocks A and B.
2 Apply the required PIs.
3 Read the resulting POs.

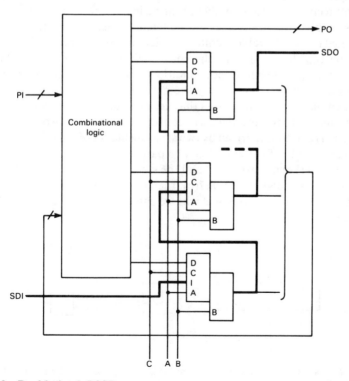

Fig. 8.6 Double-latch LSSD system.

4 Latch the remaining outputs from the combinational logic into the SRLs using a single cycle of clocks C and B.

5 Shift out the SRL contents through SDO using clocks A and B. At the same time, the state variable vector required for the next test can be shifted in.

The use of separate front-ends and separate clocks for system and scan path means that, as compared with the general SISO scheme described in section 8.2, the extra propagation delay due to the multiplexer is avoided. Against that, the requirement that every flip-flop in the system is a full master/slave driven by a two-phase clock will itself inject some additional delay in those parts of the circuit in which the operational requirement could have been met with a simpler latch.

8.3.3 Single latch LSSD

In an attempt to increase the speed of the system under normal operating conditions, a modified approach is possible. The essence of this approach is to take the state variables from L1 of the SRL instead of from L2, so that L2 is no longer in the data path.

If this technique is to be applied, bearing in mind that the L1 section of the SRL is simply a level-sensitive clocked latch, there are restrictions that must be applied to the allowable architecture. In particular, direct global feedback from the flip-flop output to the combinational logic input, as shown in Fig. 8.7(a), would result in race conditions leading to uncertain or incorrect operational behaviour. Certainly the performance of the system would be dependent on timing parameters that are difficult or impossible to control.

One solution to this problem is to consider the circuit as two blocks of flip-flops, FF(1) and FF(2), driven by two separate blocks of combinational logic, CL(1) and CL(2). If the circuit can be partitioned so that the state variable inputs to CL(1) come from FF2 and the state variable inputs to CL(2) come from FF(1), as shown in Fig. 8.7(b), then by clocking FF(1) and FF(2) with two separate non-overlapping clocks we can ensure that the race problem is avoided.

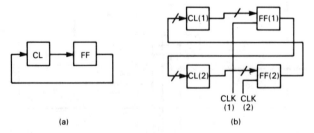

(a)   (b)

Fig. 8.7   Using level-sensitive latches. (a) Direct global feedback introduces hazards; (b) partitioning and use of separate clocks permits reliable operation.

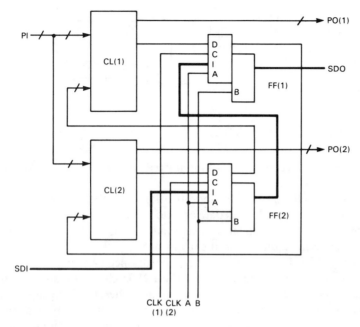

Fig. 8.8   Single-latch LSSD system.

The implementation of this principle using SRLs is shown in Fig. 8.8, where, for simplicity, a single SRL is drawn to represent each group of flip-flops, and where the scan path is again highlighted as a solid line. Because, with this system, each data pathway contains only a single latch (L1), this is known as single-latch LSSD to differentiate it from the system described in the previous section, in which each data pathway contained a double latch (L1 and L2).

The single latch system provides speed improvement at the cost of increased complexity; this includes not only the extra clock signals, extra pin-out, and extra wiring, but also the extra design effort and circuitry needed to achieve the necessary partitioning. It is also apparent that the L2 latch is now used exclusively for the testing function; assessments of the extra hardware costs (silicon overhead) to provide for testability commonly come to 20% or more with this system.

8.3.4   The L1/L2* system

The heavy silicon overhead incurred by the single latch LSSD system has encouraged a search for a less expensive alternative. The major cost component is due to the fact that the L2 latch is used only for testing, so that in total the SRL is between two and three times as complex as a simple clocked latch. If a way can be found to use the L2 latch to perform a system function,

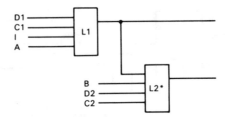

Fig. 8.9 The L1/L2* form of SRL.

then this overhead could be substantially reduced. This is the thinking behind the L1/L2* system.

The system relies on a further modification to the SRL consisting of adding a second front end to L2, with data and clock inputs, so that the modified L2, denoted by L2*, is identical in construction to L1. The complete L1/L2* latch is represented diagrammatically as in Fig. 8.9.

To use the L1/L2* SRL we require a partitioned system as with the single latch scheme. The difference is that the L2* latches, instead of being idle during systems operation, are used as system latches, receiving input through D2 and being clocked by C2. The race problem is overcome by arranging that in all cases the L1 and L2* of an individual SRL are in different partitions. The overall scheme is shown in Fig. 8.10; the scan path, as with the single latch system, threads through all the SRLs in turn, controlled by alternate activation of clocks A and B. System operation is controlled by alternate activation of clocks C1 and C2.

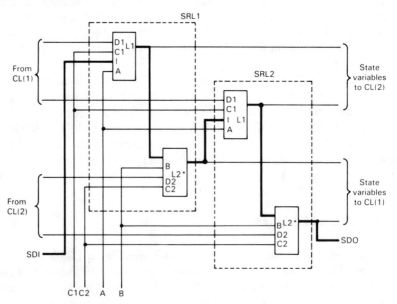

Fig. 8.10 L1/L2* implementation of LSSD.

The circuit is tested by dealing with CL(1) and CL(2) separately. To test CL(1), for example, the procedure is:

1 Shift the required state variables into the L2* latches from SDI using clocks A and B. Notice that these have to be shifted through the L1 latches, but that the values remaining in L1 are irrelevant to the test.
2 Apply the required primary inputs to CL(1) and observe the resulting primary outputs.
3 Clock the remaining outputs into the L1 latches using a single cycle of C1.
4 Shift out the contents of the L1 latches through SDO using clocks A and B. Again the relevant data from L1 is unavoidably clocked through the L2* latches on its way to the output.

Having completed this procedure for all the tests for CL(1) a similar procedure is used to test CL(2). Notice that the two cannot be done simultaneously because L1 and L2* are not independent when being used as a shift register. This is a limitation of the level-sensitive form of construction; we must operate the A and B clocks in sequence, making the SRL into a master/slave flip-flop. Hence, if the useful data is in L1 (that is, we have applied a test to CL(1)) then we must first activate clock B, which transfers the data to L2*, and then clock A, which transfers the data to the next L1. The first transfer, of course, destroys the data that was in L2*, which is why we can extract test data from CL(2) only by a separate test sequence. In this latter case, shifting out will have to start by activating clock A followed by clock B.

8.4   OTHER SCAN DESIGN METHODS

8.4.1   Scan path

As was the case with LSSD, the **scan path** system demands the exclusive use of a special flip-flop with a built-in selection mechanism enabling it to be used either for a normal system function or as an element in a shift-register chain for testing purposes.

The scan path flip-flop, whose equivalent circuit is shown in Fig. 8.11, is essentially a master/slave circuit, with two separate input systems, as with the SRL. Unlike the SRL, however, the two clocks are not used together as a two-phase system, but are separate signals which serve to select the input (either system data, D, or scan data, SDI) as well as to activate the latch. Hence, whichever mode of operation is in use, the flip-flop is controlled by a single clock, the slave section being latched through an inverter from the same clock signal that activated the master latch. This implies that the scan path flip-flop requires a less elaborate clocking system than the SRL, but the circuit

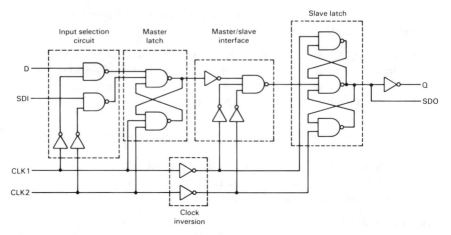

Fig. 8.11   Flip-flop used in scan path system.

contains race hazards which demand careful attention to timing if reliable operation is to be ensured.

The resting state for this flip-flop is with both clocks high. This blocks both input channels, and allows the master latch to retain data, while the slave latch is transparent and the master/slave interface is enabled; the output of the slave latch is therefore the same as that of the master. When one of the clocks goes low, the appropriate input is enabled and the master latch becomes transparent; the slave latch holds its previous value and the master/slave interface is disabled. When the clock returns high the converse sequence occurs, so that the master then retains a value corresponding to the previous input, and this value is transmitted through the interface and the slave to the output. In both phases of operation, correct behaviour depends on the latch entering its hold mode before the input circuitry is disabled; with two gate delays in the input circuit this condition should normally be met without difficulty.

It should be clear from the discussion above that the scan path system is essentially very similar to LSSD, and can be used in much the same way as the double-latch configuration. The use of a single clock rather than a two-phase arrangement will make the operation faster at the expense of abandoning the level-sensitive property and so accepting the risk of timing problems. There also seems little scope in this circuit for adopting alternative configurations to reduce the overhead or increase the speed.

8.4.2 Scan/set

As an alternative to a reconfigurable architecture in which the system flip-flops are turned into a shift register for testing purposes, a separate shift

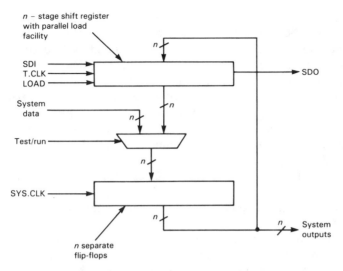

Fig. 8.12   Principle of scan/set system.

register can be provided outside the data path. This is the basis of the **scan/set** system, whose principle of operation is illustrated in Fig. 8.12. For normal system use the n separate flip-flops are driven from the system data inputs under the control of the system clock. The n-stage shift register provides a scan path from SDI to SDO under the control of the test clock (T. CLK), and it can also be parallel-loaded from the system flip-flops.

Control of the state variables is achieved by clocking the appropriate values into the shift register through the scan path, and then loading them into the system flip-flops using the system clock, having made the appropriate multiplexer selection with the test/run mode signal. Observation of state variables is achieved by parallel-loading the shift register and then shifting the results out through the scan path.

In assessing the scan/set system as compared with the other scan design methods, a number of features need to be considered:

a   The use of separate flip-flops for testing purposes suggests that the hardware overhead will be comparable to that of single-latch LSSD. There are however, compensating factors; most notably, the separation of the scan path from the system data path means that there is no race problem and so need for partitioning.

b   The discipline imposed on the designer by the scan/set system is more relaxed than with either LSSD or scan path. Not only is a special flip-flop not specified (although the incorporation of the multiplexer into the system flip-flop circuit would probably be convenient) but also, and more importantly, there is no requirement for all the system state variables to have access to the scan path. The choice of which of the system flip-flops should be connected to the scan register will be made on the basis of the controllability or observability of the various nodes. It should be noticed that it is possible to monitor through the scan path any node in the circuit, whether or not that node is the output of a flip-flop; hence, the scan/set system can incorporate an electronic equivalent of the stake pin to enhance the observability of any node, without incurring the penalties of set-up time and tester interface cost.

c If there are flip-flops in the system that are not attached to the scan register, it follows that the circuit to which test pattern generation effort will have to be applied will no longer be a purely combinational one. Even though flip-flops that are excluded will be those that are easy to test, it will still require a more sophisticated ATPG system than a purely combinational circuit would have done.

d The time required to shift data into or out of the scan register will be less than for the corresponding LSSD system because:
   i the number of stages in the shift register is less than the number of flip-flops in the system;
   ii (more importantly) the scanning process passes only through the shift register, and not through the system latches.

e Because the scan path is completely separate from the system, and because scan path and system are controlled by completely independant clocks, it is possible to perform both functions simultaneously. There are at least two ways in which this feature can be useful:
   i Test do not have to be conducted as separate entities; an initial state can be entered, the system allowed to run for some number of system clock cycles, and the result then latched into the shift register for examination.
   ii While the system is running with normal inputs, a 'snapshot' of the internal states can be taken and scanned out for inspection without interferring in any way with the system function or performance.

The important point about these facilities is that the performance is evaluated while the system is operating at normal speed, so that some measure of the dynamic performance can therefore be obtained.

The scan/set approach is not by any means the most widely used method of structured design for testability. It is, however, significant because general-purpose chips based on this approach are now available. The Serial Shadow Register device produced by AMD is an 8-bit wide device which is essentially the same as Fig. 8.12, with some additional circuitry to allow both the system data inputs and the system outputs to connect to bi-directional busses, with direct communication between the busses and the shift register (called by AMD the shadow register).

With these devices, and others that are promised, it will, for the first time, be possible to incorporate scan design economically using non-custom chips, so that the techniques will become available to system designers other than mainframe computer manufacturers.

8.4.3 Random access scan

The final structured DFT method to be described here differs from all the others in that it does not use a shift register to provide input/output access to the system latches. It shares with the other methods the objective of providing controllability and observability for all system latches, and does this by using a special latch, one form of which is shown in Fig. 8.13(a). The circuit is very similar to the first stages of Fig. 8.11, providing separate pathways for system data and test data, with system function controlled by S.CLK and 'scan' function by T.CLK. The difference lies in the address line SEL (in practice usually more than one line) which is unique to the particular latch and allows

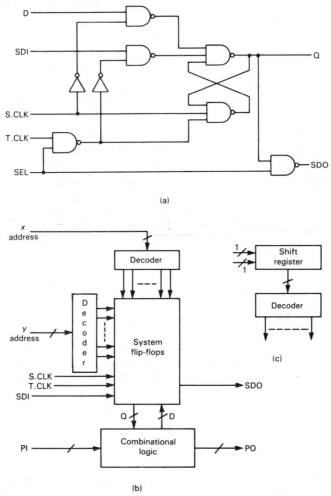

(a)

(b)

(c)

Fig. 8.13 Random access scan. (a) Addressable latch; (b) system architecture; (c) method of economizing on I/O pins.

that latch to be selected for data transfer through SDI or SDO. These I/O pins are common to all latches in the system, so that scanning data in or out consists not of shifting the data through the system, but rather of scanning through the addresses of the system latches so as to select each latch in turn. As with the scan/set system, it is also possible to observe any node within the combinational network by providing a gate with an enabling address.

The overall system is shown schematically in Fig. 8.13(b). The system flip-flops can be thought of as if they were a memory array, addressed by a two co-ordinate system. The x and y addresses can be applied to the decoders in parallel, as shown; or, if that number of input pins is not conveniently available (as will more than likely be the case), the addresses could be clocked in serially to registers as indicated in Fig. 8.13(c). This economizes on pins at the expense of test application time.

The method of testing this structure is essentially the same as for the other scan-type systems. The system is set to its required state by entering values serially into the flip-flops through SDI under the control of T.CLK and the address inputs. Similarly, the state of the system after it has been clocked can be inspected serially through SDO by scanning the addresses of the flip-flops. However, although the data is entered or inspected one bit at a time, it is not actually being shifted through the flip-flops. Thus, inspecting the state variables is a non-destructive process; the next test could proceed from that state. If there is no test requiring that state, then, since any flip-flop can be selected in any order, it is necessary to select only those flip-flops whose states need to be changed. Equally, when assessing the result of a test, there is no need to inspect every flip-flop, but only those that convey useful information. By taking advantage of this flexibility, the test pattern application time can be kept down, although test pattern generation may be made rather more complicated.

8.5 THE COST OF DFT

There is no disguising the fact that the use of structured design for testability techniques, whichever particular implementation is chosen, is an expensive business. The concern expressed at the end of Chapter 7 at the costs that can be incurred by trying to improve testability is no less appropriate in assessing DFT; it will be clear that DFT incurs cost under the same headings that were listed for *ad hoc* testability enhancement:

a additional I/O pins and consequential costs for ATE interface, and test program set-up and run time;
b additional circuitry and wiring within the circuit (often referred to, particularly with a chip, as silicon overhead);
c performance degradation and reduction in reliability due to extra components in the signal path.

In addition, all the DFT schemes require long test program application times because of the serial nature of the scan path.

In terms of I/O pins, DFT schemes have clear advantages over testability enhancement schemes, since the latter mostly require one or more pins for each node whose controllability or observability is being improved. DFT, on the other hand, provides improvement to any number of nodes with a very modest number of pins – it could be as little as one and does not normally need to be more than three or so – although minimization of extra I/O pins is always achieved at the expense of silicon overhead or test program run times.

It is not possible to make a precise general quantitative statement of the silicon overhead cost of DFT. In general, modifications for testing are largely confined to the stored-state devices; the total cost, therefore, expressed as a percentage of the cost of the circuit, will depend on the relative amounts of combinational logic and flip-flops. A rough estimation can be made by reference to the circuit diagram of the flip-flop. The SRL of Fig. 8.5(a), for example, contains 12 gates. When used in a double latch LSSD system, gates 2, 5 and 6 are used solely for testing, while all the others are in the signal path. If we assume that the combinational logic in the circuit amounts to an average of 10 gates per flip-flop, then the silicon overhead can be assessed as 3/22 or just under 14%. If the same circuit is built as a single-latch LSSD system, the L2 latch as well as the front end are testing overheads, which therefore amount to 7/22 or nearly 32%. The L1/L2* system under the same conditions, bearing in mind that the full SRL now comprises two system latches, will yield an overhead of 5/35 or just over 14%. These numbers, however, based on a simple gate count, take no account of

a the circuitry needed for partitioning,
b the cost of providing additional clock and control circuitry, or
c the chip or board area occupied by wiring.

A true figure for silicon overhead is certainly not easy to derive; the originators of LSSD talk in terms of overheads between 4 and 20%.

The other costs attached to DFT – degradation of performance and reduction of reliability – are even more difficult to quantify. Increased propagation times are of no importance in the many applications in which operating speed is not a limiting factor, and in such cases it is hardly realistic to speak of a cost. In cases where speed is important, there is clearly a choice that will have to be made between speed and testability; the choice will depend in any particular case on the priorities in the specification. The loss of reliability is even more difficult to assess, and its acceptance must be an even more subjective judgement.

The justification for all these costs comes from the economies in test pattern generation which result from being able to confine ATPG effort to purely combinational circuits without global feedback. With the continuing improvements in the technical capabilities of the fabrication process, even this simplified class of circuit can be very complex, and ATPG programs still tend to be large and slow; run-times measured in hours of CPU time (of very large mainframe computers) are commonly spoken of, and much research effort is still being expended in an attempt to find a more efficient way of generating test programs, and of making the programs themselves more efficient in terms of run-time, fault-coverage, and diagnostic capability. Until such improvements can be made, and extended to the case of the general (unstructured) circuit, it seems likely that the use of DFT techniques will be the only way that the problem can be kept within bounds.

SUMMARY
Chapter 8

The problems of testing general synchronous sequential circuits can be seen as arising from limitations on controllability and observability; in particular, the absence of direct control over the state variables makes it difficult to establish the necessary test conditions, while the absence of observability of the state variables makes it difficult to interpret the outcomes of the tests. Structural DFT addresses these problems by imposing a discipline on the circuit designer such that the tester will be able to access each flip-flop independently both of the combinational logic and of the other flip-flops. For this facility to be provided without requiring an impossible number of input/output pins, all DFT schemes make use of a scan mechanism, SISO, which requires only a few pins irrespective of the number of flip-flops accessed.

The most widely used DFT systems, including all variations of LSSD and the scan path system, make use of a reconfigurable architecture which permits the combinational logic and the flip-flops to be separated for test purposes. The flip-flops are then tested as a shift register, using a test procedure that will be essentially the same for any circuit, and the combinational logic is tested in isolation, free from all the possible complications of sequential behaviour.

Other variations of DFT achieve the effective separation of combinational logic and flip-flops without reconfiguration; either by using a auxiliary register as an intermediary, as in scan/set, or by making provision for direct access to each flip-flop one at a time, as in random-access scan. In both these cases, the system flip-flops are not reconfigured to a shift register, and read-out of the system state is non-destructive.

DFT is not achieved without cost, measured usually in terms of silicon overhead. Estimates vary typically between 4% and 20%: some of the variation in this figure arises from different methods of assessment. The justification for incurring these and other costs of DFT lies in the hope of savings in TPG and other test-related costs.

EXERCISES
Chapter 8

E8.1 Explain why a general synchronous sequential circuit often presents difficulties to the test engineer. Describe how these difficulties are overcome by incorporating the SISO approach to circuit design.

E8.2 Compare and contrast the design features of the flip-flops used in LSSD and Scan Path systems.

E8.3 What are the advantages and disadvantages of designing with an architecture that is reconfigurable for testing purposes?
What alternative approaches are possible, and what are their merits?

E8.4 Explain the difference between single latch and double latch implementations of LSSD. What factors need to be taken into account when deciding which of these design methods to adopt?

9

SELF-TESTING CIRCUITS

9.1 THE PRINCIPLE OF SELF-TESTING

9.1.1 Self-checking through redundancy

With the increasing complexity of all levels of circuitry – chip or pcb or sub-unit or system – the cost of all aspects of test activity is continually increasing. Automatic test equipment, particularly the in-circuit testers which are commonly used for large pcbs, is very expensive both to buy and to maintain; test pattern generation, even for a circuit based on scan design, can easily require a run-time of tens of hours on a large mainframe computer; and test application times, especially for diagnostic test programs, will also tend to increase exponentially with circuit complexity, resulting in a need for duplication of tester resources in order to provide the necessary throughput. It is against this background that the provision of built-in testing facilities, so that individual parts of the circuit will test themselves, has been seen to offer the hope of a cost effective alternative.

The idea that self-checking can be built into a system is not new; the use of parity bits is perhaps the simplest example of a basic system being modified in the interest of providing some measure of protection against data corruption. A single parity bit added to a data word will detect a single error, but will give no indication of which bit is wrong. Moreover, if two errors occur in a single word, that word will be classified as correct. Some features of the simple parity bit systems are worth noticing:

a   Hardware is needed to calculate the appropriate value of the parity bit and to append it to the data word; further hardware is needed to check the parity subsequently. All this hardware is entirely separate from the basic system requirement, and is, to that extent, redundant.

b   The extent of the safeguard provided is limited; in this case a single error will be detected, but double errors will not.

c As with all other schemes in which extra circuitry is built-in to the system, the overall reliability of the system will be reduced, and the operating speed may be reduced.

The need for additional hardware is inevitable in any scheme aimed at providing a facility over and above the normal system function. In all such cases the extra cost has to be justified by a reduction in costs elsewhere in the manufacturing process.

The limited nature of the safeguard provided by the simple parity scheme is also a common feature. To obtain better protection, it is possible to use more elaborate (and more expensive) schemes. An extension of simple parity, for example, is to add a parity word to each block of data such that each column in the block conforms to the required parity. A single-bit error will now generate two parity failures, one on a row and the other on a column. This enables the failing bit to be pin-pointed, so that the error can be corrected automatically. This is one form of single-error-correcting code; it will also provide detection (but not correction) of most (but not all) multiple errors. The purchase of additional protection at the expense of additional elaboration can be taken further, but the law of diminishing returns applies; the extra cost gets larger, while the extra protection gets smaller. It should also be noticed that complete protection can never be achieved; there are always some patterns of faults that would escape detection, although the probability of one of these particular patterns occurring may be considered small enough to be neglected.

The argument that incorporation of additional circuitry reduces reliability is not always valid. It is certainly true that a larger circuit will have more things to go wrong, so that the probability of a manufacturing fault being present will be higher, and the mean time between failures in operation will be lower. However, there is one special class of circuits for which additional circuitry is used to enhance overall system reliability. These are fault-tolerant circuits, which are designed in such a way that the system will continue to deliver correct outputs even though some parts of the system have failed. This is achieved essentially by functional duplication: the outputs are computed independantly by two (or, very often, three) separate systems. Failure to agree indicates the presence of a fault, and the circuitry can also be arranged to identify the most probable site of the failure. Fault-tolerant systems, by their very nature, can present a problem for testing, since a correct output does not necessarily imply a fault-free circuit. For the manufacturer to ship such a circuit with a fault already in it defeats the whole object of the fault-tolerance; to avoid this problem, the fault-tolerance will have to be disabled for testing purposes.

9.1.2 BITE: Built-in test equipment

Neither error-detection coding nor fault-tolerant design is really providing the kind of fault detection that we have been considering throughout this book. Coding is intended to guard against data corruption over transmission paths, and fault-tolerance is using massive 'redundancy' to buy ultra-reliability. Hardware fault detection has somewhat different requirements. Fault location is not normally required, since automatic correction of hardware faults is not usually possible; go/no-go testing is certainly adequate for an individual chip, and it would be a good first step for pcb or higher-level unit. The incentive to consider self-testing is the hope that this degree of fault detection can be provided without incurring the costs of automatic test equipment.

The principle of providing a circuit with built-in test facilities is illustrated in Fig. 9.1. Under normal operation, the circuit (which can be combinational or sequential) is connected through the multiplexer to its normal inputs, and the normal outputs are produced. The control and timing unit, which has a single mode control input, is responsible for providing the system clocks and for making the multiplexer selection. It can also implement reconfiguration of the circuit itself, as discussed in Chapter 7; particularly useful techniques in this context will be partitioning (see section 7.5) and breaking global feedback and other paths (see section 7.4).

In its test mode of operation, the circuit is supplied with a sequence of stored test patterns, and the outputs are compared with the stored fault-free responses; any discrepancy between fault-free responses and the observed ones will result in a 'no-go' indication.

The attractions of self-testing go beyond the saving of ATE costs. In at least two respects, self-testing can be expected to provide positive advantages from a testing point of view.

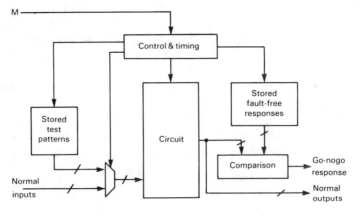

Fig. 9.1   A system with built-in-test.

a Maintenance and field-service operations will usually not have access to sophisticated ATE. If self-test is able to locate a fault to a replaceable unit (a pcb, say) detailed fault location and repair can then be carried out at a central servicing facility. This is not the same as the board-swapping procedure described in Chapter 1 (see section 1.2.3); the trial-and-error aspect is missing, so that the worst effects of the pipeline feature are avoided; the advantages of board-swapping, however, including minimal down-time and minimal skill-requirements for the field-service operative, are retained.

b Modifications to circuits to improve testability often demand additional access points for the tester (see section 7.3.1). With normal ATE, this requires extra input/output pins, which may be difficult to accommodate. A self-testing circuit can incorporate as much tester access as desired; the only additional pins needed are one for mode control and perhaps a second one for the go/no-go decision (although even this can be avoided if an indicator lamp is incorporated instead).

As it stands, the arrangement illustrated in Fig. 9.1 is not suitable for implementation. To store test patterns and responses in the circuit would require built-in ROMs with the test data programmed in; and although the comparison unit is conceptually straightforward, it would require a non-trivial amount of circuitry. All this represents hardware overhead for testing; practical implementations, particularly at chip level, seek to minimize this overhead.

9.2 PROVISION OF TEST PATTERNS

9.2.1 Stored test vectors

In instrumentation applications, a test ROM can often be accommodated without difficulty, and this approach is quite commonly used particularly in an 'intelligent' instrument, which usually incorporates one or more microprocessors, and which will necessarily include a ROM containing the operating system of the instrument. This ROM (enlarged if necessary) could also contain the test vectors.

For a self-testing chip, however, storage of test vectors in ROM is unlikely to be feasible, because it is unlikely that the necessary chip area will be available. Furthermore, as well as having to provide for storage of the vectors, the manufacturer is faced with the cost of generating the vectors in the first place. Both of these problems can be avoided with exhaustive test.

9.2.2 Exhaustive test

The chief merit of exhaustive test is that no test pattern generation effort is required; all combinations of inputs are to be applied. Equally, the test vectors do not need to be stored; they can be generated by a sequential circuit that steps through all its states.

In order to utilize an exhaustive test strategy, the circuit has to be reconfigurable for testing purposes, ensuring that the combinational logic is available separately and that it is partitioned to keep the number of inputs within bounds. Once this has been established, the only remaining problem is to build the sequence generator. The obvious way to do this is to use a standard binary counter, but it turns out that the same effect can be achieved with less hardware by taking advantage of the properties of feedback shift registers (FSRs).

Figure 9.2(a) shows a three-stage FSR with feedback connections from all three stages by way of a modulo-2 sum circuit (exclusive-OR). Analysis of this circuit reveals that its state transition diagram is as shown in Fig. 9.2(b). A slightly different FSR, with feedback from the second and third stages only, is shown in Fig. 9.2(c); its state transition diagram is shown in Fig. 9.2(d). It is clear that the form, as well as the details, of the behaviour of an FSR is dependant on the choice of feedback connections.

Any autonomous synchronous sequential circuit (that is, a sequential circuit that has no external input apart from the clock) must produce a cyclic output. This is easily demonstrated by observing that at any instant the next state must depend only on the present state. Hence, if any state is ever entered for a second time the subsequent sequence must be the same as the first; that is, the output will be cyclic. But for a circuit with r flip-flops, the number of

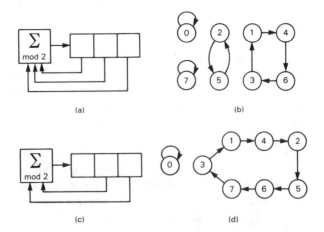

(a)

(b)

(c)

(d)

Fig. 9.2 Autonomous behaviour of LFSRs. (a) A three-stage LFSR with feedback from all stages; (b) state transition diagram for the LFSR in (a); (c) LFSR with modified feedback; (d) state transition diagram for (c).

distinct states is 2^r; so that by the time that the sequence has gone on for more than 2^r clock pulses, at least one of the states must have been entered more than once. Thus the circuit must product a cyclic output and the period cannot be more than 2^r. With an FSR a further limitation can be deduced. If the register is in the all-zeros state then the feedback quantity must be zero irrespective of what the feedback connections are. Hence, the all-zeros state must always be a 'stuck' state; the longest possible period of the shift register is therefore $2^r - 1$. If an r-stage FSR has a period of $2^r - 1$ it is said to be **maximal length**. It is clear that we cannot, with an FSR, generate a truly exhaustive test set, because of the absence of the all-zeros state, but we can come as close as possible to it by using a maximal length FSR, such as the one depicted in Fig. 9.2(c).

If an FSR is to be used to generate test patterns, the number of stages in the register will be dictated by the number of inputs to the circuit. The feedback connections will then have to be chosen so as to achieve maximal length. To understand this relationship we will analyse the behaviour of the r-stage FSR shown in Fig. 9.3, using the techniques introduced in Chapter 4 (see section 4.3.2). The first point to notice is that the sequence of values observed at the input to the shift register is also observed at every stage of the register, the only difference being that the sequence at each stage is delayed by one time interval compared with the previous stage. The performance of the FSR is therefore completely defined by the sequence generated at the input, which consists of the set of values.

$$\{a_0, a_1, a_2, \ldots, a_\infty\}$$

This sequence can be represented as a polynomial in the delay variable x as

$$S(x) = \sum_{i=0}^{\infty} a_i x^i \qquad (9.1)$$

The first value in the sequence, a_0, is computed as a function of the initial state of the register, which can be denoted as

$$\{a_{-1}, a_{-2}, \ldots, a_{-r}\}$$

and represented in polynomial form as

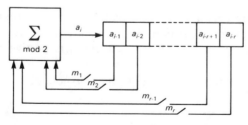

Fig. 9.3 A general r-Stage LFSR.

$$I(x) = \sum_{j=1}^{r} a_{-j} x^{-j} \tag{9.2}$$

The feedback connections, represented in Fig. 9.3 as $m_1, m_2, \ldots, m_r$, specify the terms in the sequence as

$$a_i = \sum_{j=1}^{r} m_j a_{i-j} \tag{9.3}$$

Substituting eqn (9.3) into eqn (9.1) gives

$$S(x) = \sum_{i=0}^{\infty} \sum_{j=1}^{rr} m_j a^{i-j} x^i$$

$$= \sum_{j=1}^{r} m_j x^j \sum_{i=0}^{\infty} a_{i-j} x^{i-j} \tag{9.4}$$

The second summation can be split to separate the negative and positive powers of x.

$$S(x) = \sum_{j=1}^{r} m_j x^j \left[\sum_{j=1}^{r} a_{-j} x^{-j} + \sum_{i=0}^{r} a_i x^i \right] \tag{9.5}$$

Substituting eqns (9.1) and (9.2) into eqn (9.5) gives

$$S(x) = \sum_{j=1}^{r} m_j x^j [I(x) + S(x)]$$

Rearranging, and remembering that modulo-2 subtraction is the same as modulo-2 addition, gives

$$S(x) = \frac{\sum_{j=1}^{r} m_j x^j I(x)}{1 + \sum_{j=1}^{r} m_j x^j} \tag{9.6}$$

From this equation we can make two important deductions:

a the sequence generated depends on the initial conditions, as indicated by the numerator of eqn (9.6). In particular, this verifies that an all-zero initial state will yield an all-zero sequence irrespective of the feedback connections.

b The denominator of the eqn (9.6) is a polynomial

$$P(x) = 1 + \sum_{j=1}^{r} m_j x^j$$

This is called the **characteristic polynomial** of the FSR; the form of the characteristic polynomial will clearly determine the behaviour of the FSR.

One important feature of the characteristic polynomial can be illustrated by the FSR's shown in Fig. 9.2. For the FSR of Fig. 9.2(a) we can see that the feedback equation is

$$a_i = a_{i-1} + a_{i-2} + a_{i-3}$$

so that $m_1 = m_2 = m_3 = 1$. Hence the characteristic polynomial is

$$P(x) = 1 + x + x^2 + x^3 \qquad (9.7)$$

which can be factorized to give

$$P(x) = (1 + x)(1 + x^2).$$

In the same way we can deduce that the FSR of Fig. 9.2(c) has the feedback equation

$$a_i = a_{i-2} + a_{i-3}$$

and the characteristic polynomial

$$P(x) = 1 + x^2 + x^3 \qquad (9.8)$$

This characteristic polynomial cannot be factorized; this is the essential difference between eqns (9.7) and (9.8), and constitutes a necessary condition if the FSR is to produce a maximal length sequence. An irreducible characteristic polynomial is not, however, a sufficient condition. The additional requirement also depends on factorization; if the length of the maximal length sequence is given by

$$s = 2^r - 1$$

then $P(x)$ must be a factor of $1 + x^s$, but it must not be a factor of $1 + x^t$ for $t < s$. If we think of the three stage register, for which $s = 7$, we can see that

$$1 + x^7 = (1 + x)(1 + x^2 + x^3)(1 + x + x^3) \qquad (9.9)$$

Of these three factors, the first, $1 + x$, is a factor of $1 + x^n$ for all values of n; it does not represent a possible characteristic for a FSR. The other two factors can each be shown to satisfy all three requirements:

 a they cannot be factorized further;
 b they are factors of $1 + x^7$;
 c they are not factors of $1 + x^n$ for $n < 7$.

Each of these two factors forms the characteristic polynomial for a maximal length FSR, and, in fact, these are the only two ways in which a maximal length FSR can be produced with a 3-stage shift register.

Applying these criteria in the case of a large shift register is not a trivial matter. In the last analysis, trial and error has to be employed; although there are methods of performing algebraic transformations on known maximal length characteristic polynomials to generate new ones. There are also extensive tables that have been computed and published, although these are necessarily far from exhaustive (a 16-stage register has over 2000 maximal length configurations, while a 24-stage register has over a quarter of a million). One important result that has been established is that, for a shift register of any size, maximal length can be obtained using not more than four feedback connections. This requires only three 2-input XOR gates; and it confirms the earlier statement that an FSR achieves an (almost) exhaustive coverage using much less hardware than is needed for a binary counter.

9.2.3 Random test

An exhaustive test set generated by an FSR is, in principle, very straightforward and simple to implement, but if the number of inputs is large then the time taken to apply the test sequence may be unacceptable – a set of 30 inputs has over 10^9 input combinations, so that an exhaustive test set applied at the rate of one pattern per μs would take nearly 18 minutes to complete! In these circumstances, the alternative must be to apply some much smaller number of tests. A systematically designed test set is the obvious anser, but, whether a functional or a structural approach is used, this brings back all the problems of the generation and storage of test patterns.

An alternative approach is based on the observation that, in a combinational circuit of realistic size, any particular input test vector will usually cover a significant number of faults. Any random selection of tests of reasonable size, therefore, can be expected to achieve reasonable fault-cover. If the tests are chosen on this random basis, the size of the test set needed to achieve any particular level of fault-coverage will certainly be significantly larger than that of the optimal test set; the merit of such a system, however, is that it produces a test set of manageable size while still incurring no TPG costs.

In order to make use of this principle, we need to build a sequence generator to supply the test vectors. This generator will take the form of a sequential circuit containing n flip-flops (where n is the number of inputs to the block of combinational logic to be tested), and the sequence of r states that are generated should be a random subset of the 2^n possible states of the circuit.

While the concept of a random selection of test patterns is easy to visualize, the definition of what constitutes a random sequence is very much less easy to produce. It is clear, for example, that it would not be satisfactory to use states zero to $r-1$ of a binary count sequence, because some inputs would never be exercised; our 'random' test set should at least ensure that each input takes the values 0 and 1 with approximately equal frequency through the test set.

We have already seen that a feedback shift register requires less hardware than a binary counter of the same length, and it has also become apparent that the sequence it produces is irregular, apart from the inevitable perio-dicity in the sequence as a whole. If an FSR is to be used for the generation of a random sequence of patterns, we must clearly ensure that the period of the sequence is larger than the number of test vectors to be used, since there is no point, with a combinational circuit, of applying the same vector twice. The most straightforward way of ensuring that this condition is satisfied, without putting any restrictions on the number of vectors that can be used, is to choose the feedback connections to give a maximal-length FSR.

The question of what is meant by a random sequence must be decided by postulating criteria that can be described in mathematical form. Such criteria will include the equal probabilities of 0 and 1 (mentioned above), and also restrictions on occurrences of internal patterns. A standard tool for the assessment of structure in an analogue waveform is the autocorrelation function defined by

$$R(T) = \frac{1}{T} \int_{-T/2}^{T/2} f(t)f(t - T)\mathrm{d}t \qquad (9.10)$$

A similar measure can be applied to the binary sequence entered into an FSR, although some redefinitions of the signal values are necessary. It turns out that criteria of randomness based on probability distributions and on the autocorrelation function are satisfied by the input sequence of a maximal length FSR, so that this is the method generally used to provide either exhaustive or partial random test sequences.

The biggest question raised by the use of a truncated LFSR sequence is that of effectiveness. This could be assessed by simulation, but this is a very expensive process, and its use would throw away much of the advantage of the avoidance of TPG effort. Alternatively, a guess based on the size and complexity of the circuit and the past experience of the test programmer is the best that can be used. This uncertainty over fault-cover may be one reason why BITE is viewed with some suspicion in the world of circuit design, although there have been a number of studies that suggest that, at least for some representative circuits, a relatively small number of random tests can be remarkably effective.

9.3 RESPONSE ASSESSMENT

9.3.1 Signature analysis

Having established an economical way of providing the test patterns, we should now turn our attention to the other major requirement for a self-test system – the generation of the go/no-go decision. In principle, as shown in Fig. 9.1, this requires a comparison between the actual circuit response and the fault-free response, but in practice this is usually much too expensive in

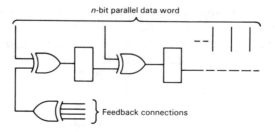

Fig. 9.4   A multiple-input signature register.

terms of hardware overhead. As with the fault location problem (see Chapter 4), the solution lies in data compression, usually based on CRC signatures (see section 4.3.2).

When examining the response of a complete circuit, we will normally have multiple outputs available, all of which need to be taken into account when assessing the result of a test. This contrasts with the diagnostic exercise, where we are usually concerned with a serial data stream appearing at a single node. One way of accomodating the more complex data array, appearing as a sequence of parallel words, is to modify the signature analysis register as shown in Fig. 9.4. The operation of this arrangement depends on the properties of the XOR gate which, as has already been established, performs modulo-2 addition. Thus if two data streams $x(t)$ and $y(t)$ are applied to the inputs of a two-input XOR gate, the output will be

$$z(t) = x(t) \oplus y(t)$$

If one input of the gate is held at 0, the gate acts as a buffer, with the output being equal to the other input. From this it follows that the output in response to two inputs applied together is equal to the sum (modulo-2) of the responses to each of the inputs applied separately. This is the classical superposition principle which, when applied to a continuous system, defines a linear system. Hence, any feedback shift register in which the feedback function is by way of XOR gates is described as a **linear feedback shift register** (LFSR).

By invoking the linearity property of the circuit of Fig. 9.4, we can see that an n-stage multiple-input LFSR can be regarded as n single-input LFSRs with each stage containing the sum (modulo-2) of the values due to each input separately. As with the single-input LFSR, the set of values in the register at the end of the test program is described as a signature.

Once the set of responses of the circuit to the test program has been collected in the form of a signature, the comparison and decision is implemented trivially. Figure 9.5 shows a four-stage register; if we suppose that the fault-free signature is 0110, the circuitry shown will generate a logic signal representing the test decision.

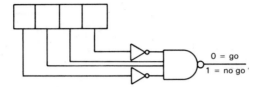

Fig. 9.5   Simple decision circuitry.

9.3.2   Sensitivity to errors

The property of linearity can be used to identify the effects of errors in the data stream on the resulting signature. Consider first, for simplicity, a single-input signature analyser supplied with a data stream

$$D = \{d_0, d_1, \ldots\ldots, d_r, \ldots\ldots\}$$

If the fault-free data stream is

$$F = \{f_0, f_1, \ldots\ldots, f_r, \ldots\ldots\}$$

we can then define an error stream

$$E = \{e_0, e_1, \ldots\ldots, e_r, \ldots\ldots\}$$

such that

$$e_r = \begin{cases} 0 & \text{if } d_r = f_r \\ 1 & \text{if } d_r \neq f_r \end{cases}$$

From this it follows that

$$D = F \oplus E$$

so that the response to D is equal to the sum of the response to F (which is, of course, the fault-free signature) and the response to E. Hence the behaviour of the LFSR can be analysed in terms of its response to the error stream.

In assessing the performance of an LFSR as a single-input signature analyser, it is, perhaps, surprising to find that the particular feedback function used has very little effect. This is because of the general form of the circuit (a shift register) which ensures that any given state has the same two successor states and the same two predecessor states, irrespective of what the feedback connections may be.

Consider, for example, the three-stage LFSR shown in Fig. 9.6(a), whose full state transition diagram is shown in Fig. 9.6(b). We can readily see that by removing all the '1' transitions we will derive the state transition diagram of Fig. 9.2(d), which represents the autonomous behaviour of the same LFSR. Less obviously, by simply removing appropriate links from the state transition diagram of Fig. 9.6(b), we can derive the state transition diagram of Fig. 9.2(b), which represents the autonomous behaviour of an LFSR with

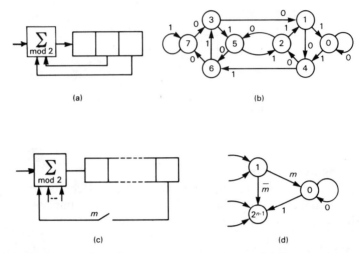

Fig. 9.6   Properties of single-input signature register. (a) A particular register;
(b) state transition diagram for (a); (c) a general n-stage register;
(d) part transition diagram for (c).

different feedback connections. In short, every three-stage shift register, whatever its feedback connections (or even if it has no feedback connections at all), has a transition diagram of the shape shown in Fig. 9.6(b). The only difference that the feedback function makes is to interchange the ones and zeros on some of the transition conditions. For example, a general n-stage register as shown in Fig. 9.6(c), in which m represents the presence or absence of a feedback connection from the last stage, will have transitions among states 0, 1, and 2^{n-1} defined as in Fig. 9.6(d).

The performance of the signature analyser can be deduced from the state transition diagram by considering the response of the system to an error stream, E. The feature of interest is the final state of the system; in particular we are interested in those error streams that leave the system in state 0, which means that the final signature is the same as the fault-free one. If E consists of all zeros, this is the correct response to a fault-free data stream; if E contains any ones, this represents undetected errors. The number of undetected error sequences, therefore, is simply the number of ways of traversing the diagram and finishing in state 0. Hence, since the state transition diagram for a given size shift register is always the same shape, whatever particular feedback connections are used, it follows that the only difference that a change of feedback function can make is in which particular sequences escape detection; the total number of such sequences is not changed.

A data stream containing a single error corresponds to an error stream E consisting of a single one followed by an indefinite succession of zeros. In this situation, we can see from Fig. 9.6(d) that the one in E will take the system out of the state 0, and that if $m = 1$ then we can guarantee that the system cannot re-enter state 0. If $m = 0$, therefore, the possibility of re-entering state

0 will depend on the possible ways of entering state 1. The predecessor states are states 2 and 3; if the transition condition from either of these states to state 1 is to be zero there must be no feedback connection from stage $m - 1$. By following this analysis all the way back to the first stage, we can deduce that, provided there is at least one connection (from any stage), a single error will always be detected; this confirms the observations made previously (see section 4.4.4).

If E contains multiple ones then the situation is less clear out. It is apparent (considering, for example, the transition diagram of Fig. 9.6(b)) that there will be sequences other than the error-free one that will leave the system in state 0, and that these error-containing sequences will therefore fail to be detected. (This feature is sometimes referred to as **aliasing**.) It is, however, also apparent that the number of incorrect sequences that escape detection is only a small fraction of the total number of possible incorrect sequences. This again is in accordance with the results deduced in section 4.4.4.

The feedback connections used with a single-input signature analyser do not affect the overall error-detection performance. This conclusion needs to be modified when we come to analyse the performance of the **multiple-input signature register** (MISR). As pointed out in the previous section, the linearity property ensures that this behaves as though it were a set of n single-input registers with the outputs added together to form a single composite signature. However, the individual equivalent registers have their single inputs fed in at various points along the chain. The effect of this can be illustrated by the equivalent circuit of Fig. 9.7(a), which shows a three-stage MISR with feedback from all stages, and all external inputs zero apart from the one to the second stage. The transition diagram for this circuit is shown in

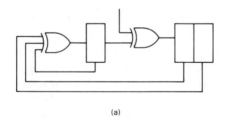

(a)

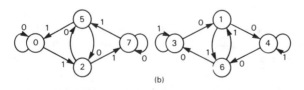

(b)

Fig. 9.7   Analysis of an MISR. (a) Equivalent circuit with input to second stage active and all others held at 0; (b) state transition diagram for (a).

Fig. 9.7(b); because the states are in two separate groups, it is clear that not all of the possible signatures can ever in fact be generated. As a result, the proportion of error sequences that will go undetected will be larger than for a normal single input register (in this particular case, one in four rather than one in eight). This characteristic is clearly undesirable.

If the MISR is supplied with all zeros on the data inputs, the circuit is equivalent to an autonomous LFSR, since all the XOR gates act simply as buffers. If, therefore, the feedback connections are chosen so as to make the register maximal length, this will ensure that all the states are attainable, and hence that the error detection performance is not degraded. Furthermore, the same considerations will apply to each of the multiple inputs, so that the same set of feedback connections will provide optimal error detection performance for each of the incoming data streams. With this in mind it is not hard to see that, as with the simple signature register, a single error occurring anywhere in any of the data streams is certain to result in a signature that differs from the fault-free one, and hence is certain to be detected. By following through essentially the same analysis as in section 4.4.4, we can also deduce that, for an n-stage register, the probability that an incorrect data set will give rise to the fault-free signature tends to 2^{-n} for long test vector sequences.

9.4 BUILT-IN SELF TEST

9.4.1 BILBO

Just as the principle of scan design has led to a number of design methodologies intended to ensure that every design is testable (see Chapter 8), design methodologies to support **built-in self test** (BIST) are beginning to appear. One such scheme, which has been extensively quoted, is the **Built-in-Logic Block Observer** (BILBO). The core of this system is the BILBO register, which consists essentially of a number of basic cells connected together as shown in Fig. 9.8. Each cell consists of a synchronous D-type flip-flop together with supporting logic. In addition to the data inputs and outputs for each flip-flop, $D_1 - D_n$ and $Q_1 - Q_n$, the BILBO has an additional input, SDI, and two mode controls B_1 and B_2. These mode controls allow the system to be configured in four different ways.

1 $B_1 = 0, B_2 = 1$
For each stage of the register, gates 1 and 2 are both disabled, so that the input XOR gate, gate 3, has both its inputs forced to zero irrespective of the values of the data inputs or the contents of the preceding stage. Hence one clock pulse while in this mode will reset all stages to zero. This provides for initialization without the need for any asynchronous inputs to the flip-flops.

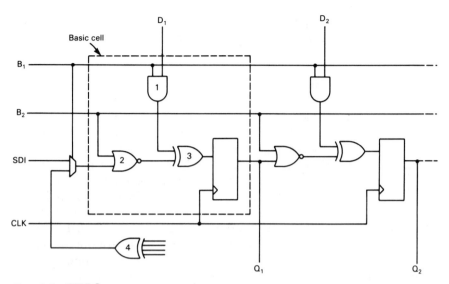

Fig. 9.8   BILBO.

2 $B_1 = 1, B_2 = 1$

Gate 2 is disabled and gate 1 is enabled. Hence the links between stages are all broken, and the register acts as n separate flip-flops each loaded from its own data input. This is the normal system operation mode.

3 $B_1 = 0, B_2 = 0$

Gate 2 is enabled and gate 1 is disabled. Hence each stage is connected to the next stage effectively through an inverter. The paths from the data inputs are blocked, but the path from SDI is open, so that the whole device has now become a shift register which can be used to implement scan path operations, the SDO output being available at Q_n. A BILBO in this mode could also be used to implement a shift register in normal system operation; in this case access to the $\overline{Q}$ outputs would allow difficulties due to the inter-stage inversions to be removed.

4 $B_1 = 1, B_2 = 0$

With gates 1 and 2 both enabled, and with the multiplexer feeding the output of gate 4 into the first stage of the register, the circuit is configured as an LFSR. The feedback connections to the input of gate 4 are chosen in accordance with the conditions discussed in section 9.2.1 so as to produce a maximal length register. In this mode of operation, the register can perform two functions.

   a With the data inputs fed from the outputs of a circuit undergoing test, the BILBO acts as an MISR.

   b If the data inputs are held at a constant value, the BILBO will generate a maximal length pseudo-random sequence, provided that the input values are chosen so as to avoid the 'lock-up' state. Because of the inversions between stages, the lock-up value for the data inputs is all ones. Hence, it is acceptable for the register to be initialized to the all-zeros state.

9.4.2 Methods of use

The way in which the BILBO principle can be used to implement built-in test will depend in detail on the particular system to which it is applied. A bus-oriented system, for example, will usually divide naturally into separate modules as shown in Fig. 9.9. These modules will often require output registers for operational reasons; otherwise special ones can be added. Either way, if these registers, marked B in Fig. 9.9, are implemented as BILBOs an effective self-testing structure results. The BILBOs themselves can be tested by way of the scan path. Initialization can be achieved using the BILBO reset, or alternatively by using the scan path to initialize to any arbitrary state. The internal circuitry of modules Y and Z can then be tested by putting BILBO X into its random sequence mode to provide the test inputs through the data bus, and putting BILBOs Y and Z into their signature analysis modes to collect the test results. At the end of the test sequence the signatures are available for inspection through the scan path. Module X can then be tested by putting BILBO X into signature analysis mode, and using Y or Z to provide the random test sequences.

In practice, a system will seldom be quite as simple as the picture shown in Fig. 9.9 suggests. In addition to the data bus inputs, modules will frequently have mode controls, which, from the testing point of view, are effectively additional circuit inputs that need to be treated on the same basis as the data inputs. Furthermore, the data outputs of any particular module are not necessarily sufficient to provide satisfactory observability of the internal activity of the module. These problems can be overcome essentially by adding

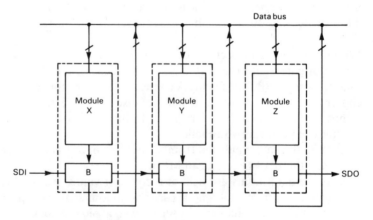

Fig. 9.9   Using BILBO in a bus-structured system.

extra stages to the registers to provide for the testing requirements; these extra stages would be inoperative during normal system use, being brought into play for testing by the use of multiplexers.

9.5 OBTAINING FAULT-FREE SIGNATURES

Having established an architecture that will support self-test, we are left with a system that produces a set of signatures, and we may feel reasonably confident that a faulty system will produce signatures that differ from the fault-free ones. The problem now is to discover what the fault-free signatures are. This is a fundamental problem not only in self-testing circuits, but also for the application of any testing procedure whether for go/no-go or diagnostic testing, and with or without the use of ATE.

There are two ways of tackling this problem. The first is the theoretical approach – calculating the answers from a knowledge of the circuit. In practice the only way this can be done for a circuit of realistic size is to use simulation. This is, in fact, the solution most commonly adopted, but it does have a major drawback. Simulation is a very costly process; a circuit of fairly modest complexity (say, 5000 gates) will require computer time measured in hours. Developments in fabrication technology will soon make 100 000 gate chips commonplace, and wafer scale integration will produce still more complex systems for analysis. Full simulation to establish the results of millions of tests being applied to a circuit of this complexity may well be impossible to perform economically, especially for relatively small volume production parts.

The alternative approach is an experimental one. This requires a **known good chip** (KGC) or **known good board** (KGB), which can be exercised by the test program, and whose resulting signatures can be collected by an external test system. This method is used with some board testers to produce a diagnostic data base by collecting good signatures from internal nodes with the aid of a guided probe. The scheme works well if once the KGB or KGC has been identified; the trouble is that the only way of being certain that the part is good is to test it! The circularity in this argument is unavoidable, and it means that if signature collection relies on this approach, there must inevitably be some residual possibility that those signatures are wrong.

There are two experimental procedures that have been used to collect signatures. The first is applicable when a prototype system is available into which the chip or board is intended to fit, and consists of two stages.

    a A short test pattern sequence for the part is generated using manual methods. The sequence may consist of one or two hundred tests for a 5000-gate circuit, and can be expected to cover perhaps 60% of the single-stuck faults.

b Parts that satisfy the short test are put into the prototype system, and the system which is verified, using a short functional test.

A part that satisfies both the above tests is then regarded as the KGC or KGB, and the full test sequence is then applied to determine the required signatures.

The second experimental procedure, which is really the only one possible if no prototype system is available, is very simple. It consists of taking a batch of prototype parts, and measuring the signatures actually obtained in response to the test program. If two or more such parts give the same signature, then that signature is taken to be correct.

9.6 THE COST AND EFFECTIVENESS OF SELF TESTING

There is no doubt that the incorporation of BIST into a system, whether at chip, board, or higher functional level, brings with it significant costs to the manufacturer. As with all design disciplines that require concessions to testing requirements, these costs can be categorized under five headings.

a **Additional hardware costs.** One or more additional pins are required on each unit. Extra circuitry is needed to support reconfiguration. In many BIST implementations, separate registers are added to the basic circuit to provide TPG and signature analysis. Control circuitry must be provided to apply the test and to assess the results. Finally, all these additional features require additional wiring: within a chip, this can consume significant silicon area; even on a board, or in the back-plane wiring between boards, interconnections can make a significant contribution to overall costs.

b **Increased manufacturing costs.** A physically larger system will cost more to build. In the case of an individual chip, this increased cost will be manifested by a reduction of yield.

c **Decreased reliability.** A circuit with more components can be expected to have a smaller MTBF.

d **Performance degradation.** Additional components will consume power, and, if they appear within signal pathways, they will increase propagation delays.

e **Design cost.** A larger circuit will take longer to design, and will incur larger costs in computer time for design verification and for layout and routing.

These production costs are partly offset by the removal of TPG cost (which for a high-density chip may well run into hundreds of hours of CPU time) and the reduction of ATE requirements. The latter is not just a financial incentive; current and projected complexity of circuits require high volumes of test pattern data, and this, together with continually increasing operational speeds, makes demands on the capabilities of ATE that are increasingly difficult to satisfy.

The justification for incorporating BIST has to be in terms of reductions in lifetime cost of the product. The chief hopes of economic advantage with BIST are concerned with maintenance and repair. Ideally it would be possible to achieve accurate and rapid fault detection, with very short MTTR (mean time to repair) and this should be achievable with the use of low-skill personnel and without the need for expensive ATE. In practice it has not always fulfilled this promise. The central problem is the false-alarm rate – the BIST system diagnoses faults in units that subsequently appear to be fault-free. The reasons for this are not entirely clear, and, indeed, the definition of what constitutes a false alarm is open to question. However, it is clear that in the present state of the art, BIST, is not proving to be the answer to all the problems of maintenance and repair, and that further research work is needed to identify the causes of the present difficulties and to establish definitive BIST strategies for optimum performance. Nevertheless, it seems clear that as the overall complexity of systems continues to increase, while the observability of internal nodes continues to decrease, the use of BIST techniques is bound to become more and more widespread.

SUMMARY
Chapter 9

A major impetus behind the wish to incorporate BITE into systems is to simplify field service and maintenance. Taken together with possible savings in TPG costs and in the complexity of ATE needed for initial production testing, this represents an attractive proposition.

The simplest implementation of this idea, in which a test program derived in conventional ways is stored in a ROM as part of the system, is used in some instrumentation applications; for general system use, however, this approach has severe limitations, and the main interest lies in a method in which the test program is generated rather than being stored. An exhaustive test has further attractions in that no TPG effort is required, and it can be implemented most conveniently by using a maximal length LFSR.

The theory of the LFSR demonstrates the importance of the characteristic polynomial and, in particular, serves to bring out the relationship between the pattern of feedback connections and the length of the sequence produced.

Exhaustive test can be contemplated only if the circuit

Summary
Chapter 9
continued

has a relatively small number of inputs, or can be partitioned into sub-circuits of a suitable size. In other cases the built-in test principle can be implemented using a random subset of the possible input patterns. These can also be generated with an LFSR, preferably of maximal length. The effectiveness of such a random sequence is necessarily a matter of conjecture, although the indications are that good results can be obtained.

The decision mechanism for BITE invariably uses data compression so that the only comparison necessary is with a single signature. The effectiveness of signature analysis, particularly when using an MISR, can be assessed by reference to the theory of the LFSR; by using a maximal-length pattern of feedback connections single-error sequences are certain to be detected, and the probability of failing to detect multiple-error sequences can be made arbitrarily small by increasing the size of the register.

All the principles discussed above are brought together in BILBO, which is a practical implementation that has found some favour and many imitators.

EXERCISES
Chapter 9

E9.1 A conventional 5-stage LFSR is in state 1 0 0 1 1.

    a What are the predecessor states?
    b What are the successor states?

Would these results be different if

    c the feedback were non-linear;
    d the feedback went to intermediate stages?

E9.2 What are the characteristic polynomials of the two LFSRs shown?

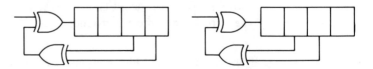

Determine (algebraically) whether either or both of the registers are maximal length. Confirm your conclusions by working out the state sequences (starting from an initial all 1s state) in response to a fixed 0 at the input.

E9.3 A certain ROM is provided with an additional bit in each word whose value is chosen to give even parity across the word, and an additional word whose bits are chosen to give odd parity across the corresponding bits in every word. Find an example of a multiple error in the ROM which will not be detected by examination of the check bits.

E9.4 A three-stage LFSR with feedback from the second and third stages only is maximal length. Show that with respect to a single input applied to the second stage (as in Fig. 9.7a) the register can enter all eight states.

E9.5 Outline the essential features of a BILBO register, and describe the various modes of operation that are available.

E9.6 Explain how a system can be configured using BILBO registers to implement built-in test. Discuss the problem of evaluating the fault-free signatures in such a system.

10

FURTHER READING

This book has set out to give an introduction to the subject of testing digital circuits, and, in the limited space available, it has been possible to give no more than an introduction.

The literature of the subject is very large, and is growing more and more rapidly. This chapter will seek merely to suggest some ways in which this literature may be found. The references have been selected to include recent review papers, which will, among other things, direct attention to the classical literature, and to avoid, as far as possible, sources such as conference proceedings and internal company reports to which most students will not have ready access. The references are introduced in a sequence corresponding broadly to the order in which topics have been dealt with chapter by chapter in the body of the book, although inevitably many papers overlap chapter boundaries.

There are very few textbooks devoted exclusively or primarily to testing. Breuer and Friedman (1976) present a thorough coverage of the subject with a strongly mathematical bias. Bennetts (1982) presents a much more pragmatic view. The latter is, arguably, the only realistic approach to circuits of the complexity to be found in today's devices, since logical manipulations of functions of many variables consume unacceptable amounts of computer time.

CHAPTER 1

The economic factors about testing strategy are discussed in great detail by Davis (1982), who ably demonstrates the extraordinary complexity of the arguments that have to be considered. Test economics provides continuing material for discussion at the annual IEEE International Test Conference, and is never far from the surface in any work in this field. The calculations are never simple; Bowers and Pratt (1985), for example, seek to demonstrate how buying cheap ATE turns out more expensive in the end.

An exposition of the characteristics of typical commercial ATE systems is given by Stover (1984). Such material is bound, by its nature, to be obsolescent in terms of detail as the various ATE manufacturers continually vie with each other to develop machines with better capabilities than those of their competitors, but the underlying principles of the systems do not really change very much.

Some of the general difficulties confronted by the test engineer, such as the differences in detail in the performance of different versions of nominally the same component, are described by Waters (1982); similar comments are also made by Bennetts (1982). This is a problem that particularly confronts the system designer who is concerned with seeing consistent performance of his system. A second source is important to ensure continuity of supply, but the attendant difficulties then have to be faced.

CHAPTER 2

Structural TPG depends crucially on fault-models, so that it is not surprising that a great deal of attention has been, and continues to be, directed to this topic. A review by Hayes (1985) summarizes briefly but clearly some of the current thinking, including a description of the MOS faults that are not well represented by the stuck-fault model. This problem has, not surprisingly, attracted a good deal of attention in recent years; Hayes (1984), Burgess *et al.* (1985), and Jain and Agrawal (1985b) are a few of the many workers who have sought alternative models to provide a more satisfactory basis for TPG in MOS circuits.

All fault-models have their critics; there is no doubt that the simplified pictures presented in the body of this book can all be shown to be inadequate. Spencer and Savir (1985) have considered some of the finer points of fan-out faults, and Kodandapani and Pradhan (1980) found problems with bridging faults. Nevertheless, as pointed out by Williams (1984), the single-stuck-fault model continues to dominate the practice of testing because it has consistently been found that a test program that covers a high percentage (say, more than 95%) of these faults gives an acceptable detection rate for real defects. A good review of the whole field of testing is given by Muehldorf and Savkar (1981).

CHAPTER 3

Automatic (computer-based) TPG is a whole subject on its own; all methods
are based ultimately on the sensitive path principle although detailed ways of
implementing this vary greatly. The classical method is to use the *D*-
algorithm, described by Roth (1980) (and originated by him), and also in
some detail by Bennetts (1982). Modified versions that are said to be more
computationally efficient have been developed: a very detailed description of
one of these (PODEM/RAPS) is to be found in Bennetts (1984). Assessment
of fault-cover can be achieved with a fault-simulator; alternatives, which are
said to be cheaper in computer time, are described by Abramovici *et al.* (1984)
and Jain and Agrawal (1985a).

CHAPTER 4

Signature analysis is widely used as a diagnostic tool, and has proved remarkably effective, although doubts about the theoretical assessment of its performance have been expressed by Smith (1980). These doubts centre on the
assumption of equal probability of occurrence of the possible error
sequences; however, no evidence seems to have been produced to suggest that
practical performance falls significantly below the theoretical ideal.

CHAPTER 5

The improved representation of the operation of a JK flip-flop (equations
5.2) was suggested by Bennetts (1984). Testing sequential circuits is so
difficult, as this chapter attempted to demonstrate, that the need for
structured DFT techniques is almost universally accepted. Methods for
deriving test patterns for sequential circuits using extensions of the *D*-
algorithm are described by Breuer and Friedman (1976), but the emphasis
nowadays is on avoiding the problem rather than trying to solve it.

CHAPTER 6

As might be expected, a great deal of attention has been paid in recent years to
the problems of testing specific LSI structures. This attention is largely
focused on the formulation of appropriate fault-models; a review by Gai *et
al.* (1983) summarizes the current position.

Bose and Abraham (1982) have concluded that the fault set developed to model cross-point defects also accounts for defects in other parts of the PLA structure, and have described a method of deriving minimal test sets to cover them. This method is based on a knowledge of the product terms, but there has been considerable effort devoted to test schemes that are independent of the particular logic function implemented. Treuer *et al.* (1985) review the field and present a method of their own.

The fault-effects experienced in RAMs have been the subject of an enormous literature, and new variations of test procedures continue to appear. The algorithm described in section 6.4.3(d) was proposed by Nair *et al.* (1978). Among the more recent schemes are those of Saluja and Kinoshita (1985) and Papachristou and Sahgal (1985). The particular problem of soft errors due to alpha-particles was described by May and Woods (1979), and has been reviewed by Carter and Wilkins (1986).

The importance of the microprocessor, together with its obvious complexity, have meant that test procedures for this class of device have received considerable attention. Daniels and Bruce (1985), in an interesting and entertaining paper, have given a historical account of the changes at Motorola from their earliest microprocessors, in which testing was viewed as an afterthought, to the current position, in which testing is built in to the structure. For users other than the microprocessor manufacturers, the main emphasis has (necessarily) been on functional testing; typical examples of this approach are those of Thatte and Abraham (1980) (the S-graph), Robach and Saucier (1980), and Bellon *et al.* (1982).

CHAPTER 7

Many measures of testability have been devised as an aid to circuit improvement or as a tool in ATPG. Bennetts (1984) gives a very detailed description of one such measure (CAMELOT) and briefly reviews a number of others. Other testability measures and their uses in TPG are described by Hess (1982), Kirkland and Flores (1983), and Chen and Breuer (1984).

General guidelines about good design practice as regards testability (apart from structured DFT techniques) have appeared from time to time over the past couple of decades. Comprehensive discussion of the topic is to be found in Bennetts (1982, 1984). A recent paper of interest from this point of view is by Roberts and Lala (1984) who discuss a formal method of identifying optimal ways of partitioning circuits with minimal overhead.

CHAPTER 8

Structured DFT based on SISO is widely accepted, and in some cases, such as the semi-custom gate array UK 5000 described by Cosgrove (1985), the CAD system prevents the designer from adopting any other design structure. The general field of structured DFT has been reviewed with great thoroughness by Williams and Parker (1982) and Williams (1984).

A comparative study of several alternative methods of reconfiguring a system for test purposes was presented by Buehler and Sievers (1982), but they were unable to demonstrate that any one method showed clear superiority.

CHAPTER 9

There seems little doubt that built-in-self-test techniques will become increasingly applied particularly to VLSI chips, and the publications on this subject are proliferating. Recent reviews have been written by Maunder (1985) and McClusky (1985), and the subject has also been included in the general DFT reviews listed under Chapter 8.

The BILBO technique, first described by Könemann et al. (1979), has been extensively quoted. Komonytsky (1983), Fasang (1983), Beucler and Manner (1984), and LeBlanc (1984) are some among many who have implemented schemes based essentially on this principle.

The impossibility of performing exhaustive test on a circuit with a large number of inputs has led to an interest in applying a truncated random sequence. Experimental results with such sequences have been reported by Williams et al. (1985) and Williams (1985); the effectiveness of very short sequences is remarkable.

A variation on the simple BILBO, used by Beucler and Manner (1984) and also included in the study by Williams et al. (1985), uses the signature at each clock time as the test stimulus for the following clock time, thus removing the need for a separate test pattern generator. This also seems to provide remarkably good fault-cover.

Gate arrays are becoming increasingly important in semi-custom design. As with Cosgrave (1985), the device described by Resnick (1983) has a built-in structure that ensures the existence of a scan path and permits BILBO-like operation. The whole field of BIT for gate arrays has been reviewed by Totton (1985).

Finally, a new system devised by Plessey has been described in very general terms by Smith (1985). This structure is used within a CAD system which automatically adds the additional components and pathways necessary for self-testing to the basic system specified by the designer. The scheme is based on a 'structured test register', which has been described by Paraskeva et al. (1985), and which is essentially an extended form of BILBO, with additional operating modes. The hardware overhead is said to be about 20%, which is the same order of cost as other DFT structures.

CHAPTER 10

Having read all the papers listed here, and the references contained in those papers, the student will be up to date – as at 1985! To keep up to date, he needs to scan the current literature. The most important sources are:

IEEE Design and Test
IEEE Transactions on Computers
Proceedings of International Test Conference

Sources that often include papers on various aspects of testing include:

Proceedings of Design Automation Conference
Proceedings of Fault-Tolerant Computing Conference
IEE Proceedings, Parts E and G
IEEE Computer
VLSI Design

REFERENCES

Abramovici, M., Menon, P.R. and Miller, D.T. (1984). Critical path tracing: an alternative to fault simulation. *IEEE Design & Test* **1** (1), 83–93.

Bellon, C., Liothin, A., Sadier, S., Saucier, G., Velazco, R., Grillot, F. and Issenman, M. (1982). Automatic generation of microprocessor test programs. *Proc. 19th Design Automation Conf.*, 566–73.

Bennetts, R.G. (1982). *Introduction to Digital Board Testing*. Crane-Russak, New York.

Bennetts, R.G. (1984). *Design of Testable Logic Circuits*. Addison-Wesley, London.

Beucler, F.P. and Manner, M.J. (1984). HILDO: The highly integrated logic device observer. *VLSI Design* **V** (6), 88–96.

Bose, P. and Abraham, J.A. (1982). Test generation for progammable logic arrays. *Proc. 19th Design Automation Conf.*, 574–80.

Bowers, G.H. and Pratt B.G. (1985). Low-cost testers: Are they really low cost? *IEEE Design and Test* **2** (3), 20–28.

Breuer, M.A. and Friedman, A.D. (1976). *Diagnosis and Reliable Design of Digital Systems*. Computer Science Press, New York.

Buehler, M.G. and Sievers, M.W. (1982). Off-line, built-in test techniques for VLSI circuits. *Computer* **15** (6), 69–82.

Burgess, N., Damper, R.I., Shaw, S.J. and Wilkins, D.R.J. (1985). Faults and fault-effects in NMOS circuits – impact on design for testability. *Proc. IEEE* **132G**, 82–9.

Carter, P.M. and Wilkins, B.R. (1986). Alpha-particle-induced soft errors in NMOS RAMs: a review. Submitted to *Proc. IEE, Section I*.

Chen, T.H. and Breuer, M.A. (1985). Automatic design for testability via testability measures. *Trans IEEE* **CAD4**, 3–11.

Cosgrove, B. (1985). The UK 5000 array. *Proc. IEE* **132G**, 90–92.

Daniels, R.G. and Bruce, W.C. (1985). Built-in self-test trends in Motorola microprocessors. *IEEE Design and Test* **2** (2), 64–71.

Davis, B. (1982). *The Economics of Automatic Test*. McGraw-Hill, London.

Fasang, P.P. (1983). Microbit brings self-testing on board complex microcomputers. *Electronics* 56, 116–19.

Gai, S., Mezzalama, M. and Prinetto, P. (1983). A review of fault models for LSI/VLSI devices. *Software & Microsystems* 2, 44–53.

Hayes, J.P. (1984). Fault modelling for digital MOS integrated circuits. *Trans IEEE* CAD3, 200–7.

Hayes, J.P. (1985). Fault modelling. *IEEE Design and Test* 2 (2), 88–95.

Hess, R.D. (1982). Testability analysis: an alternative to structured design for testability. *VLSI Design* III (2), 22–9.

Jain, S.K. and Agrawal, V.D. (1985a). Statistical fault analysis. *IEEE Design and Test* 2 (1), 38–44.

Jain, S.K. and Agrawal, V.D. (1985b). Modelling and test generation algorithms for MOS circuits. *Trans IEEE* C34, 426–33.

Kirkland, T. and Flores, V. (1983). Software checks testability and generates tests of VLSI design. *Electronics* 56, 120–24.

Kodandapani, K.L. and Pradhan, D.K. (1980). Undetectability of bridging faults and validity of stuck-at fault test sets. *Trans IEEE* C29, 55–9.

Komonytsky, D. (1983). Synthesis of techniques creates complete system self-test. *Electronics* 56, 110–15.

Könemann, B., Mucha, J. and Zwiehoff, G. (1979). Built-in logic block observation techniques. *Proc. IEEE Test Conf.*, 37–41.

LeBlanc, J.J. (1984). LOCST: a built-in self-test technique. *IEEE Design and Test* 1 (4), 45–52.

Maunder, C. (1985). Built-in test: a review. *Electron and Power* 31, 204–8.

May, T.C. and Woods, M.H. (1979). Alpha-particle-induced soft errors in dynamic memories. *Trans. IEEE* ED26, 2–9.

McCluskey, E.J. (1985). Built-in self test techniques. *IEEE Design and Test* 2 (2), 21–8.

Muehldorf, E.C. and Savkar, A.D. (1981). LSI logic testing: an overview. *Trans. IEEE* C30, 1–16.

Nair, R., Thatte, S.M. and Abraham, J.A. (1978). Efficient algorithms for testing semiconductor random access memories. *Trans. IEEE* C27, 572–6.

Papachristou, C.A. and Sahgal, N.B. (1985). An improved method for detecting functional faults in semiconductor random access memories. *Trans IEEE* C34, 110–16.

Paraskeva, M., Knight, W.L. and Burrows, D.F. (1985). New test structure for VLSI self-test: the structured test register (STR). *Electronics Letters* 21, 856–7.

Resnick, D.R. (1983). Testability and maintainability with a new 6K gate array. *VLSI Design* IV (2), 34–8.

Robach, C. and Saucier, G. (1980). Microprocessor functional testing. *Proc. IEEE Test Conf.* 433–43.

Roberts, M.W. and Lala, P.K. (1984). An algorithm for the partitioning of logic circuits. *Proc. IEE* 131E 113–18.

Roth, J.P. (1980). *Computer Logic, Testing, and Verification*. Computer Science Pres/Pitman, New York and London.

Saluja, K.K. and Kinoshita, K. (1985). Test pattern generation for API faults in RAM. *Trans IEEE* C34, 284–7.

Smith, J.E. (1980). Measures of the effectiveness of fault signature analysis. *Trans IEEE* C29, 510–14.

Smith, K. (1985). Plessey custom chips will test themselves. *Electronics Week*, March 25, 17–18.

Spencer, T.H. and Savir, J. (1985). Layout influences testability. *Trans IEEE* C34, 287–90.

Stover, A.C (1984). *ATE: Automatic Test Equipment*. McGraw-Hill, New York.

Thatte, S.M. and Abraham, J.A. (1980). Test generation for microprocessors. *Trans IEEE* **C29**, 429–41.

Totton, KAE (1985). Review of built-in test methodologies for gate arrays. *Proc. IEE* **132E**, 121–9.

Treuer, R., Fujiwara, H. and Agarwal, V.K. (1985). Implementing a built-in self test PLA design. *IEEE Design and Test* **2** (2), 37–48.

Waters, D.G.P. (1982). The problems of testing large scale integrated circuits. *Brit. Telecom. Engng* **1**, 64–9.

Williams, T.W. (1984). VLSI testing. *Computer* **17** (10), 126–36.

Williams, T.W. (1985). Test length in a self-testing environment. *IEEE Design and Test* **2** (2), 59–63.

Williams, T.W. and Parker K.P. (1982). Design for testability – a survey. *Trans IEEE* **C31**, 2–15.

Williams, T.W., Walther, R.G., Bottorff, P.S. and DasGupta, S. (1985). Experiment to investigate self-testing techniques in VLSI. *Proc. IEE* **132G**, 105–7.

SOLUTIONS TO EXERCISES

The notes that follow are intended to give an indication of the more important points raised by the exercise questions. In most cases, there is no unique solution to the question: the answer suggested here should be viewed as one possible approach, which is not guaranteed to be the best!

CHAPTER 1

E1.1 The benefits of Goods Inwards Testing are summarized in section 1.2.2. The costs depend on the thoroughness of the test (which determines the complexity of the ATE required – see section 1.3.4) and are justified by economic considerations (see section 1.4.1).

E1.2 **a** See section 1.3.1.
         b See section 1.3.3.

E1.3 See section 1.2.3.

E1.4 See section 1.3.3.

E1.5 See section 1.3.2.

CHAPTER 2

E2.1 **b** The absence of static hazard can be demonstrated by timing analysis using the constant gate delay model.
         c All single-stuck faults are covered by
               {000/0; 001/1; 110/1; 111/0}

E2.2 See sections 2.1.3 and 2.1.4.

E2.3 **a** 5.
         b 4 (assuming the four NAND gate implementation).

E2.4 The logical function of the circuit is
$$Z = A + \bar{B} + C.D + \bar{E}$$
This can be realized if the circuit given is modified by removing the inverter and the OR gate. This realisation has no undetectable faults. Other redesigns can also achieve the same result.

E2.5 **a** All faults covered except $A/0$, $C/0$, $E/1$.
  b $A.\bar{B}.C/\bar{Z}$ covers $A/0$, $C/0$.
   $A.B.\bar{C}/\bar{Z}$ covers $E/1$.
  Fan-out branch faults not covered are
  i A input to NAND gate s-a-1.
  ii D input to NAND gate s-a-1.

E2.6 **a** $\{1000/1; 1010/1; 0010/1; 0011/1\}$.
  b $\{1100/0; 1110/0; 0110/0; 0111/0\}$.
  The tests differ only in the values of B and the output. This is bound to be so because of the nature of a sensitive path.

E2.7 **a** Short-circuiting track at a to Earth will be modelled by *trunk* s-a-0.
     Tested by $1110/1$.
  b Bridging fault: high will be pulled low.
     Tested by $1100/0$.
  c Modelled as $S/1$.
     Undetectable.
  Circuit needs to be redesigned to remove redundancy.

E2.8 A valid test is a set of input values for which the fault-free output, Z, and the faulty output, Z', are different. Thus $Z \oplus Z' = 1$.
  But if X is s-a-1, $Z' = f$.
  Hence set of tests is given by $(X.f + \bar{X}.g) \oplus f = 1$ which gives $\bar{X}(f \oplus g) = 1$.
  Similarly for X s-a-0.
  For a circuit with 30 inputs the computation required to evaluate these expressions will almost certainly be unacceptable.

E2.9 See sections 2.1.4 and 2.5.
  (Other fault-models will be introduced in later chapters: see sections 3.4, 6.2.2, 6.5.4.)

CHAPTER 3

E3.1 There is no fault-equivalence or fault-dominance for an XOR gate.
  Fault-matrix (inputs: A,B; Output; Z)

	$A/0$	$A/1$	$B/0$	$B/1$	$Z/0$	$Z/1$
t_0		✓		✓		✓
t_1		✓	✓		✓	
t_2	✓			✓	✓	
t_3	✓			✓		✓

E3.2 **a** 14.
  This circuit is heavily redundant ($R = C.D$). It contains a large number of undetectable faults. It is an object lesson on the effects of bad design!

E3.3 **a** $\{A/0; D/0; F/0\}$ $\{C/0; E/0; G/0\}$ $\{F/1; G/1; H/1\}$.
b and **c**. These demonstrate that not all cases of indistinguishability can be deduced from the topology alone.
E3.4 F_1 and F_4.

CHAPTER 4

E4.1 A minimum of four tests will cover all single-stuck faults: the choices are given by
$$(t_6.t_{15} + t_7.t_{14})(t_4.t_{10} + t_{11}.t_{12})$$
There are four sets of indistinguishable faults:
$$\{A/0; E/1; G/1; J/0\} \{B/0; C/0; F/1\} \{D/1; H/1; K/0\} \{A/1; E/0\}.$$
The last three sets each consist of a group of faults associated with a single gate; only in the first group are two gates confused.
E4.2 See sections 4.3, 4.4.1 and 4.4.4.
E4.3 See section 4.1.
E4.4 $Q = 110010$; $R = 111$.
E4.5 As in Fig. 4.5, with $r = 4$; $m_0 = m_2 = m_4 = 1$; $m_1 = m_3 = 0$.
E4.6 **a** 0110.
    b New signature is 1110.
    c 100000001.
E4.7 **a** 1.56%
    b 1.47%
    c 0.

CHAPTER 5

E5.1 The 7473 is a master/slave, whereas the 74LS73 is edge-triggered.
E5.2 **a** Consider effect of each pin fault, and choose tests to check it.
    e.g. To test for C_D s-a-1 we need $Q = 1$, then $C_D = 0$. To obtain $Q = 1$, initialise with $C_D = 0$ (this does not test C_D s-a-1) followed by $C_D = 1$; then $D = 1$; then apply CP; then $C_D = 0$.
    Similarly for other pin-faults.
    b The function of the device could be described by
      i in synchronous mode ($C_D = 1$) changes of D should not affect Q while CP is unchanged (at either value). With CP, Q follows D (both senses).
      ii When $C_D = 0$, $Q = 0$ irrespective of D and CP.
E5.3 Unless the flip-flop can be initialized, it is not possible to define its behaviour.
    a For initial $Q = 0$, make $T = 1$ then apply CP. This allows $Q = D$ (for Q s-a-0).
    b For initial $Q = 1$ and $T = D$: apply CP to give $Q = \bar{D}$.
    c A T flip-flop cannot block transmission or produce fixed values except by use of asynchronous inputs.
E5.4 **a** By making $A = B = C = 0$ we produce 1 at each D input irrespective of the states of the flip-flops. One CP will then force $X = Y = 1$.

b Because of the limited access to the inputs and (especially) the outputs of the flip-flops, a hybrid strategy (see section 5.4.1) seems likely to be difficult to implement. A structural strategy based on pin-faults may be the most suitable.

Consideration should also be given to the construction of the flip-flop: in many cases, if 'Q' output is s-a-0, the '$\overline{Q}$' output will be s-a-1.

c The following is one possible set of test patterns. Each test consists of the sequence

1 Set inputs
2 Apply CP
3 Observe output

Test	ABC	Z	Test	ABC	Z	Test	ABC	Z
1	000	0	4	000	0	7	001	1
2	100	1	5	011	1	8	011	0
3	100	0	6	000	0	9	001	0

Notice that when assessing fault-cover it is necessary to check for any particular fault that

i it produces a change at one of the D-inputs before CP;
ii the resulting change in X or Y after CP is transmitted to the output.

E5.5 First make $A = 1$. Then if $W = 0$, $(VW)^+ = 01$; while if $W = 1$, $(VW)^+ = 10$. Then make $A = 0$. For $(VW) = 01$ or 10, $(VW)^+ = 00$. A more straightforward procedure would be possible with an extra PI which could force $K_v = K_w = 1$.

E5.6 a The circuit cannot be initialized because there is no way of defining the state of W.

b The approach used in **E5.4** seems most likely to succeed, noting that the initialization input can be used within the test sequence where convenient. Notice that correct operation of the initialization facility itself also needs to be verified.

It will be found that this particular circuit, even with the initialization facility, is very difficult to test: long sequences are needed to establish the states necessary to set up the sensitive paths, and the derivation of these sequences relies largely on intuition.

The experience of wrestling with this problem should convince the student that the problem is a real one!

CHAPTER 6

E6.1 Fan-out faults (see section 3.4) and bridging faults (see section 2.5) may not be covered. Also defects internal to the multiplexers could manifest themselves as other than stuck-pin faults.

a Dry joint at A will be represented by s-a-1 branch fault. It will be covered by 110/0, which also covers eight other faults.

b Short circuit will be trunk fault (since there is no break in the track). It is also covered by 110/0.

c Bridging fault can be covered by 111/1 (also by several alternatives).

E6.2 Arithmetic circuits are usually responsive to any random pattern, since a single bit change anywhere is likely to change the output.

Pick first (randomly) 00000/000. It will be found that this covers every node s-a-1.

Next try 11111/111. This covers s-a-0 faults on all except four nodes.

All remaining faults can be covered by one of several tests, e.g. 10101/001.

E6.3   **a** With any two inputs at 1, $C_0 = 1$ irrespective of the value at the third input $(0, 1, D, \bar{D})$.

Any two inputs at 0 will produce $C_0 = 0$.

It is not possible to block fault transmission to S.

  b Any two inputs at 1 will establish $C_0 = D$, and any two at 0 will establish $C_0 = \bar{D}$. The third input in each case is immaterial.

To establish $S = D$ requires exactly one or three inputs at 1; for $S = \bar{D}$, two or zero inputs at 1 are required.

  c To transmit D (or $\bar{D}$) at one input, make the other two inputs 0: this gives $S = D$ (or $\bar{D}$). If the other two inputs are 10, $C_0 = D$ (or $\bar{D}$).

E6.4 See section 6.2.2.

E6.5   **a** $\bar{A}.\bar{C}; B; A.\bar{B}; \bar{A}.\bar{B}.C.$

  b $Y = \bar{A} + B$ (after reduction).

$Z = A + B$ (after reduction).

  c **i** If the A connection is missing, $P_3 \rightarrow \bar{B}$.

    ii If an extra B connection appears on the first product term, $P_1 \rightarrow \bar{A}.B.\bar{C}$.

    iii An extra output connection between P_4 and Z.

    iv A missing connection between P_4 and Y.

  d **i** $ABC = 000, Z = 0 \rightarrow 1$

    ii $ABC = 000, Y = 1 \rightarrow 0$

    iii $ABC = 001, Z = 0 \rightarrow 1$

    iv $ABC = 001, Y = 1 \rightarrow 0$

E6.6 Any decoder defect can only cause incorrect selection of ROM location. If (by chance) the incorrect selection results in the correct outputs, the fault will be undetectable. Otherwise, the exhaustive test will cover it. There is therefore no need to make any special provision for decoder faults.

E6.7 A minimal test devised on this basis might be:

Select location 00. . . .00; Write 0; Read 0.

Select location 11. . . .11; Write 1; Read 1.

The coverage of such a procedure is clearly almost non-existent!

E6.8 If we write to and read from the same location, we have no way of telling which location has been addressed. The MARCH algorithm verifies that previous data is undisturbed before entering a write/read cycle.

E6.9 The microprocessor operates by accessing the location whose address is set up in the program counter, reading the contents of the location (which will be interpreted as an instruction) on its data lines, and incrementing the program counter. If the instruction is NOP, the cycle is simply repeated.

Hence, by fixing the data inputs at the values corresponding to NOP, the program counter will increment continuously, the result appearing at the address lines.

CHAPTER 7

E7.1 See section 7.1.2 (and also many places earlier in the book: these concepts are fundamental to the whole business of TPG).

E7.2 See section 7.2.1, together with the earlier discussion in section 2.4.

E7.3 See section 7.3.1.

E7.4 See section 7.3.2.

E7.5 The circuit can be modified along the lines of Fig. 7.8 (see section 7.5). Two additional edge-connector pins are needed to provide control of the two pairs of multiplexers. These are available, but notice that there are not enough pins available to bring out all the internal nodes as PI or PO.

    An exhaustive test of the original circuit requires 2^{20} (over a million) test patterns; the modified circuit requires $2^{14} + 2^{12} = 20480$ patterns. Taking the 74158 as a model, a 4-bit 2 – 1 MUX contains 15 gates, and a 2-bit 2 – 1 MUX would contain 9 gates. The modification requires two of each: hence overhead is about 5%.

CHAPTER 8

E8.1 The problems outlined in section 8.1 are really a re-statement of the problems discussed in Chapter 5 (see section 5.4.5 and example E5.6). SISO is described in section 8.2.

E8.2 See sections 8.3.1 and 8.4.1.

E8.3 The alternatives to a reconfigurable structure involve increased numbers of I/0 pins, either at the edge-connector or with stake-pins (see section 7.3.1).

E8.4 See sections 8.3.2 and 8.3.3.

CHAPTER 9

E9.1 **a** 00110; 00111.
    b 01001; 11001.
    c No.
    d Yes.

E9.2 **a** $1 + x^3 + x^4$.   **b** $1 + x^3 + x^3$.
    a is maximal length,   **b** is not.

E9.3 Two errors in one word will give correct parity for that word, but will generate two column parity errors. If another word also has two errors in the same bit-positions, then all parity checks will be satisfied.

E9.4 If the register is in state 0 and has an input of 0, it will remain in state 0, but an input of 1 will take it to state 2. If it is in any state other than the stuck one, then, since it is maximal length, it will cycle through the seven non-stuck states in response to a constant 0. It can enter state 0 if it receives an input of 1 while in state 4. Hence, wherever it starts, it can enter all eight states.

E9.5 See section 9.4.1.

E9.6 See sections 9.4.2 and 9.5.

INDEX

undetectable 33, 38, 89, 125
See also Fault model
Fault blocking 32, 50
Fault collapsing 45, 57, 105
Fault cover 27, 29, 85, 100, 105, 167
Fault dictionary
See Diagnosis
Fault-free output 23, 29, 50, 175
See also Signature analysis: fault-free
signature
Fault list 26, 45, 49, 80, 88, 97, 105
Fault location
See Diagnosis
Fault masking 26
Fault matrix 43, 57
Fault-model 26, 181
bridging 37, 181
cross-point 104
See also PLA
fan-out 48, 181
flip-flop 80
MOS 181
single-stuck 26, 29, 38, 45, 56, 61, 80, 97,
110, 117, 125, 181
Fault ordering 49, 51
Fault propagation
See Fault transmission
Fault site 29, 49
Fault tolerance 159
Fault transmission 29, 33, 50
Feedback
disabling 129
global 22, 72, 78, 92, 129, 146, 160
local 22
Feedback shift register (FSR)
characteristic polynomial 165
linear feedback shift register (LFSR) 67,
69, 107, 162
maximal length 163, 165
random sequence 166, 184
stuck (lock-up) state 163, 173
Field service 9, 10, 92, 161
Fixed value 28, 31, 49
Flip-flop
asynchronous inputs 79, 80, 130, 132
edge-triggered 77
fault-model 80
level sensitive 77, 143, 146
master–slave 77, 143
mathematical model 77
test sequence 80
Functional testing
See Test pattern generation

GALPAT 111
Glitch 23, 35, 78, 125
Global feedback
See Feedback
Go–nogo 55, 160, 167
Goods inwards testing 7, 16
Growth fault
See PLA
Guarding 16
Guided probe
See Diagnosis

Hazard
See Glitch

Hybrid testing
See Test pattern generation

Image 49
In-circuit testing 16
Indistinguishable fault
See Fault
Initial conditions 80, 86, 164
Initialization 80, 86, 93, 132
Input/output (I/O) pins 5, 57, 139, 140
Instruction set 24, 116
Integrated circuit (ic)
defects 8, 16, 23, 25, 37
drop-ins 5
production testing 3, 14, 16, 25, 55
See also Goods inwards testing
test procedures 4, 5, 7, 56
Internal nodes 21, 30, 59, 99, 123, 128, 135,
152, 154, 177

Jumper 130

Known good board (KBG) 175

Level-sensitive flip-flops
See Flip-flops
Level-sensitive scan design (LSSD) 77, 143,
151
double latch 145
L1/L2* latch 147
shift register latch 143
single latch 146
LFSR
See Feedback shift register
Local feedback
See Feedback
Lock-up state
See Feedback shift register
Logic tester
See Automatic test equipment
Loop breaking 72, 129

MARCH 112
Master–slave flip-flop
See Flip-flops
Maximal length
See Feedback shift register
Memory 12
Abraham's algorithm 113
pattern sensitivity 111
RAM 108, 135, 183
ROM 106
soft errors 111
Memory behind the pins 12
Microprocessor 24, 102, 114, 135, 183
Minimal test set 43
MISR
See Signature analysis
Mod-2 arithmetic 63
implementation in hardware 65
See also LFSR
polynomial representation 64, 65
remainder 64, 67
residue 67, 69
Monostable 135
Multiple faults 26, 105
Multiple input signature register (MISR)
See Signature analysis
Multiplexer
test sequence 97